MULTIPHASE FLOW IN POROUS MEDIA

Edited by

P. M. ADLER

Institut de Physique du Globe de Paris, France

Reprinted from
Transport in Porous Media
Vol. 20, Nos. 1 & 2 (1995)

SPRINGER-SCIENCE+BUSINESS MEDIA, B.V.

Library of Congress Cataloging-in-Publication Data

A C.I.P. Catalogue record for this book is available from the Library of Congress.

ISBN 978-90-481-4645-1 ISBN 978-94-017-2372-5 (eBook)
DOI 10.1007/978-94-017-2372-5

Printed on acid-free paper

Table of Contents

Multiphase Flow in Porous Media – Preface

Multiphase flow in porous media is a field of great industrial interest, but it still requires more research before it can be properly understood. This double issue of *TiPM* is devoted to some of the aspects of this fascinating topic.

Since this field is very large, we have tried to be specific and most contributions deal with logic aspects of multiphase flow and their modeling. The issue is roughly divided into three parts.

The first one is devoted to lattice Boltzmann techniques which appear to be promising and the various contributors illustrate some of their aspects.

The second part deals with three studies at the pore scale where the phenomena are addressed in more physical terms than in the previous part. The basic idea is to incorporate the results into large-scale numerical models. Two of the contributions illustrate the very important case of three-phase flow.

The last part directly addresses the upscaling of single-phase flow by a renormalization technique which is very efficient.

We may conclude that if we cannot make any claim of completeness in this crucial field, we believe that some of its recent developments are significantly addressed here.

PIERRE M. ADLER

Transport in Porous Media **20:** 3–20, 1995.

Lattice-Boltzmann Simulations of Flow Through Fontainebleau Sandstone

BRUNO FERRÉOL* and DANIEL H. ROTHMAN
Department of Earth, Atmospheric, and Planetary Sciences, Massachusetts Institute of Technology, Cambridge, MA 02139, U.S.A.

(Received: May 1994)

Abstract. We report preliminary results from simulations of single-phase and two-phase flow through three-dimensional tomographic reconstructions of Fontainebleau sandstone. The simulations are performed with the lattice-Boltzmann method, a variant of lattice-gas cellular-automation models of fluid mechanics. Simulations of single-phase flow on a sample of linear size 0.2 cm yield a calculated permeability in the range 1.0–1.5 darcys, depending on direction, which compares qualitatively well with a laboratory measurement of 1.3 darcys on a sample approximately an order of magnitude larger. The sensitivity of permeability calculations to sample size, grid resolution, and choice of model parameters is quantified empirically. We also present a qualitative study of immiscible two-phase flow in a sample of linear size 0.05 cm; simulations of both drainage and imbibition are presented.

Key words: lattice-Boltzmann simulations, single-phase flow, two-phase flow, Fontainebleau sandstone.

1. Introduction

What is the microscopic basis of macroscopic transport in porous media? How do macroscopic transport properties depend on size? These are two of the many questions that have dominated fundamental studies of single-phase and multiphase flow through microscopically-disordered materials such as porous sedimentary rocks.

Such questions may be addressed in a variety of ways (Thompson *et al.*, 1987; Cushman, 1990; Adler, 1992). Perhaps the most direct method is to conduct a series of laboratory measurements on samples of varying sizes and types. Alternatively, in the quest for theoretical understanding, one may seek to better understand the mathematical or physical basis of the Darcy-scale equations for macroscopic transport. Lastly, one can consider studies based on numerical simulation.

Each of these approaches has strengths and weaknesses. Laboratory measurements are of undisputed value as the markers of 'ground truth'; they may, however, require great expense, and are usually limited to sample sizes not much bigger than a core. Theoretical studies, on the other hand, are not necessarily limited to a specific scale, but they usually require simplifying assumptions (e.g., periodic geometries

---

* Current address: Elf Aquitaine, Département Techniques et Specialités, Division Gisement, CSTJF, Avenue Larribeau, 64000 Pau, France.

or simple scaling laws) that have only a partial relevance to reality. Numerical simulation usually attempts to bridge the gap between theory and experiment. It is typically hampered, however, by either the need to simplify geometry, physics, or both, as in the case of network models (Chandler *et al.*, 1982; Koplik and Lasseter, 1985; Dias and Payatakes, 1986a,b).

In recent years, however, another approach to the numerical study of fluid transport through porous media has gained some popularity. The idea is to numerically solve the Navier–Stokes or Stokes equations in a realistic microscopically disordered geometry, and then study how volume-averaged properties of the flow relate to microscopic details of the geometry, the flow, or both. Compared to network models, such studies offer the ability to study the micro-physical basis of macroscopic transport without the need for simplified geometries or physics; they are however limited to samples of small size (i.e., about ten pores on each side of a cube). For the case of single-phase flow, there have been direct solutions of the Stokes equations by finite-difference methods (Adler, 1992; Martys and Garboczi, 1992; Schwartz *et al.*, 1993, 1994; Martys *et al.*, Spanne *et al.*, 1994). Another approach to single-phase flow simulations has been the application of lattice-gas (Rothman, 1988; Chen *et al.*, 1991; Kohring, 1991) and related lattice-Boltzmann (Cancelliere *et al.*, 1990) methods. For the case of multiphase flow, the only numerical simulations of porous flow (at the level of the Stokes or Navier–Stokes equations) have been achieved by lattice-gas and lattice-Boltzmann methods (Rothman, 1990; Gunstensen and Rothman, 1993; Soll *et al.*, 1994).

This paper is devoted to the numerical study of single-phase and two-phase flow through a tomographically-reconstructed three-dimensional image of Fontainebleau sandstone. The simulations are performed by a lattice-Boltzmann method. Since lattice-gas and lattice-Boltzmann methods are still unconventional, we begin our discussion with a brief review of these methods in addition to providing some motivation for their use. We then describe the digital volume through which we simulate the flow. Our present results are only preliminary. We provide a study of the permeability of a small piece of rock, approximately 0.2 cm on a side, including a study of how permeability varies as a function of scale. Our results for the permeability of the largest sample compare qualitatively well to both laboratory measurements and previous calculations of single-phase flow by a finite-difference method (Schwartz *et al.*, 1994). They thus partly serve to validate the application of the lattice-Boltzmann method to porous flow, in addition to providing a detailed empirical study of how permeability varies with size at scales smaller than a centimeter and within the same piece of rock. Finally, we illustrate our capability to simulate two-phase flow through the rock, and point the way toward further studies.

2. The Lattice-Boltzmann Method

Extensive reviews of the lattice-gas method, and the lattice-Boltzmann method that derives from it, are now available (Benzi *et al.*, 1992; Rothman and Zaleski, 1994). We thus provide only the barest details below.

The lattice-gas method, originally due to Frisch, Hasslacher, and Pomeau (Frisch *et al.*, 1986, 1987), consists of populating a regular lattice with particles that hop from site to site in discrete time steps and collide with each other in a way that conserves mass (i.e., particle number) and momentum. The microscopic dynamics of such a model of a fluid are simple, but the macroscopic dynamics contain essentially the same complexity as hydrodynamics itself. Specifically, the macroscopic behavior of single-phase lattice-gas models asymptotically solves the incompressible Navier–Stokes equations:

$$\nabla \cdot \mathbf{u} = 0, \tag{1}$$

$$\partial_t \mathbf{u} + (\mathbf{u} \cdot \nabla)\mathbf{u} = -\frac{1}{\rho}\nabla p + \nu \nabla^2 \mathbf{u}. \tag{2}$$

Here $\mathbf{u}$ denotes the velocity, p the pressure, ν the kinematic shear viscosity, and ρ the density. The first equation arises from the conservation of mass in the lattice gas. The second equation derives from the conservation of momentum. The fact that it is isotropic, whereas the underlying lattice is anisotropic, is one of the remarkable features of the lattice-gas method. Usually, to achieve isotropy, 2D simulations are performed on a hexagonal lattice, and, due to a mathematical quirk, 3D simulations are performed on a 4D face-centered hypercubic lattice that is projected down to 3D.

The utility of lattice-gas methods for studying flow through porous media is now reasonably well established (Rothman, 1988; Chen *et al.*, 1991; Kohring, 1991). There are basically two motivations for such applications. First, the no-slip boundary condition of hydrodynamics is easily implemented as a simple 'bounce-back' reflection at solid walls, thus making simulations of flow through complex geometries no more difficult than simulations of open flows. Second, the discrete nature of the lattice-gas method makes it computationally efficient in terms of the work necessary to update a single site of the lattice. (For the lattice gas, a site update is usually achieved by one or more table-lookup operations, whereas classical methods require many floating-point operations.) Although lattice-gas calculations can be quite noisy, studies of macroscopic transport require only averages over the entire volume. Thus, the noise is averaged away, allowing retention of the original gain in efficiency.

For applications to porous flow the lattice-Boltzmann method is probably less efficient than the lattice-gas method, but in 3D it is easier to implement. There are now a variety of lattice-Boltzmann models in the literature (see Benzi *et al.* (1992) for a review). For this study, we use the method known as the 'relaxation-

Boltzmann' scheme, or, less descriptively, the 'BGK' method (Qian *et al.*, 1992; Chen *et al.*, 1992).

In the lattice-Boltzmann method one works with real-valued mean particle populations rather than discrete particles. As in the lattice gas, these mean particle populations hop from site to site on a regular lattice, and are redistributed at each time step by a collision operator. In the relaxation-Boltzmann scheme, the mathematical expression for collisions and propagation is extraordinarily simple. Specifically, the following evolution equation governs the dynamics:

$$N_i(\mathbf{x} + \mathbf{c}_i, t + 1) = (1 - \omega)N_i(\mathbf{x}, t) + \omega N_i^{eq}(\mathbf{x}, t). \tag{3}$$

Here $N_i(\mathbf{x}, t)$ is the mean particle population located at lattice site $\mathbf{x}$ at discrete time step t and moving with velocity $\mathbf{c}_i$ towards the neighboring lattice site located at $\mathbf{x} + \mathbf{c}_i$. N_i^{eq} is the 'equilibrium' mean population; it is a specified function of the local density $\rho = \Sigma_i N_i$ and the local velocity $\mathbf{u} = \rho^{-1}\Sigma_i \mathbf{c}_i N_i$. The parameter ω is the relaxation parameter; it determines transport properties of the resulting fluid. In particular, the kinematic viscosity is given by

$$\nu = \frac{1}{6}\left(\frac{2}{\omega} - 1\right). \tag{4}$$

To be linearly stable, one must have $0 < \omega < 2$. We note that our 3D simulations are performed on a cubic lattice (i.e., a 3D projection of the face-centered hypercubic lattice) in which the velocities make connection with the 18 nearest neighbors in addition to allowing a population at rest (the so-called $D3Q19$ model (Qian *et al.*, 1992)).

The introduction of two or more immiscible fluids follows the scheme presented by Gunstensen *et al.* (Gunstensen *et al.* 1991; Gunstensen and Rothman, 1992) and applied here to the relaxation-Boltzmann model. The two fluids are represented by two distinct populations, typically referred to as 'red' and 'blue'. The evolution equation then includes an extra term that redistributes the particle populations in a way that conserves mass, momentum, and color, but also creates an anistropy in the local pressure tensor such that the pressure in the direction of the local color gradient is increased while the pressure in the direction perpendicular to the color gradient (and therefore parallel to an interface) is decreased. From such a scheme surface tension can then be deduced.

3. Microtomographic Image of Fontainebleau Sandstone

Figure 1 is a three-dimensional digital image of the pore space of a small cube of Fontainebleau sandstone of linear dimension 0.048 cm. This image is constructed by using advanced X-ray microtomography combined with high-intensity synchrotron X-ray sources (Flannery *et al.*, 1987). Each voxel of this 3D image has a linear dimension of roughly 7.5 μm and takes on the value 1 (solid) or 0 (void); the image itself contains 64 voxels on a side.

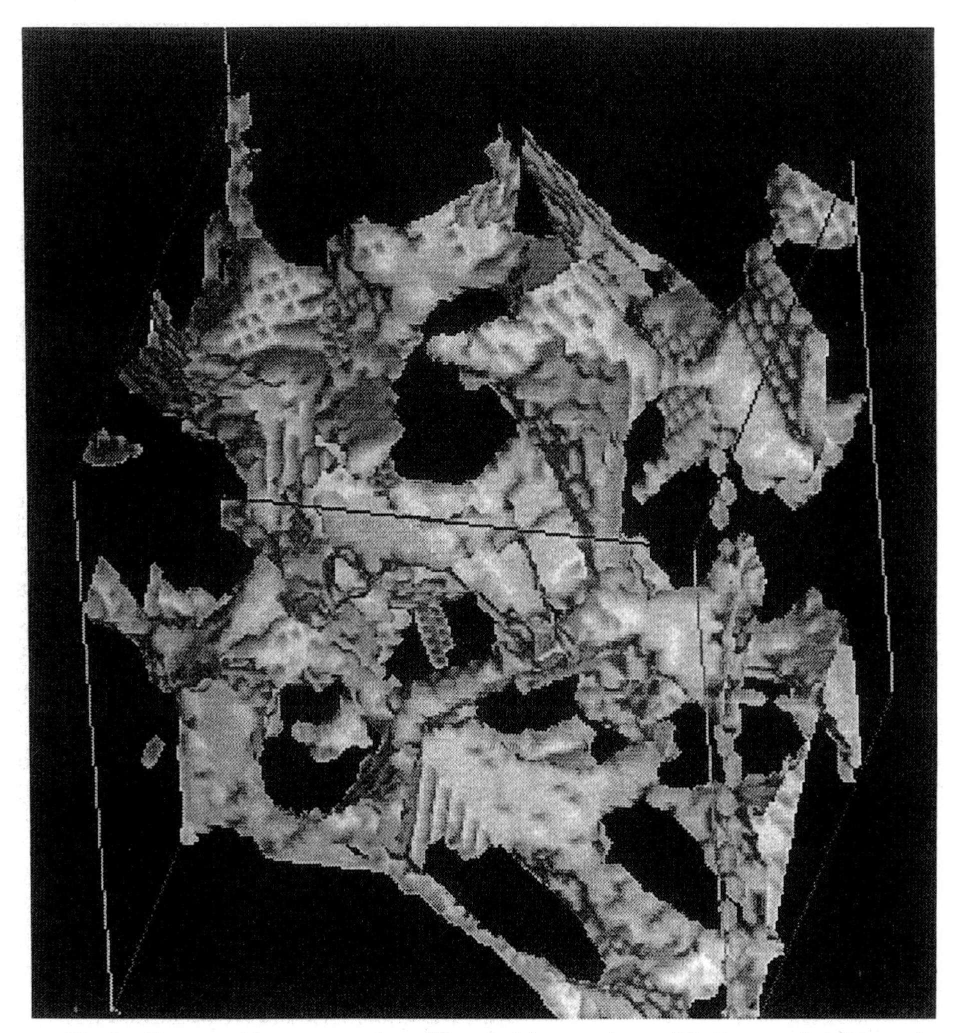

Fig. 1. Tomographic reconstruction of Fontainebleau sandstone. The linear scale is 64 voxels, or 0.048 cm, on a side. The pore space is shown in gray; the volume fraction of the pore space is about 0.15.

Some aspects concerning the construction of the binary image from tomographic gray-scale data are given by Schwartz *et al.* (1994). Remarkably, the porosity of the tomographically-reconstructed rock (with about 200 voxels, 0.15 cm or 7 grains on a side) is 14.84%, which is within 2.6% of the experimentally obtained porosity of the real rock, 14.47% (Schwartz *et al.*, 1994). Schwartz *et al.* also compare calculated and experimentally obtained values of the formation factor, permeability, and the ratio of pore volume to surface area. Of particular interest here is the permeability: using a finite-difference method, Schwartz *et al.* calculated a permeability of 1.0 darcy, which compares qualitatively well to the experimentally measured permeability, $k_{\mathrm{exp}} = 1.3$ darcys, of a cylindrical sample of roughly 4 cm in length and 2 cm in diameter.

4. Lattice-Boltzmann Simulations of Single-Phase Flow

In this section, we first illustrate pictorially one of our simulations of single-phase flow, in addition to providing a few technical details concerning boundary conditions and the measurement of permeability. Then we discuss how and why permeability measurements can depend on the resolution of the underlying grid. Finally we provide a detailed study of how permeability varies with scale in a small sample of Fontainebleau sandstone.

4.1. ILLUSTRATION OF SINGLE-PHASE FLOW

Figure 2 shows flow through the small (64^3 voxels) sample shown in Figure 1. One sees that the flow is very fast through the smallest throat in the main channel.

A few words concerning the execution of this simulation, and the others similar to it, are worth mentioning.

The ideal simulation of such a flow would be with periodic boundary conditions in each of the three orthogonal directions. To achieve this, one would duplicate the rock with mirror symmetries in each direction, and the computational mesh would then be eight times larger. To avoid that extra expense, we have instead chosen to 'jacket' the rock with walls on the four faces parallel to the overall flow direction and to add 10 layers of void space to one of the faces perpendicular to the overall flow. The medium is then made periodic in the flow direction.

In order to make the fluid flow, we simulate a body force by adding a fixed amount of momentum in the α direction, δq_α, to each of the lattice sites within the void space at each time step. When the steady state is reached, we compute the permeability in the α direction, k_α, from Darcy's law:

$$k_\alpha = \left(\frac{a}{M}\right)^2 \frac{\phi\nu\Sigma q_\alpha}{\Sigma\delta q_\alpha}. \tag{5}$$

Here ϕ is the porosity of the sample (i.e., the volme fraction of void sites), the sums are performed over all the void sites within the sample, and $q_\alpha = \Sigma_i c_{i\alpha} N_i$ is the α-component of momentum at a site. (Implicitly, $c_{i\alpha}$ is the α-component of velocity $\mathbf{c}_i$, and the sum over i is performed over all the velocity vectors $\mathbf{c}_i$ that connect a site at location $\mathbf{x}$ to its neighbors at $\mathbf{x} + \mathbf{c}_i$.) The first term converts model units into darcys. The factor a is the linear size, in microns, of a voxel in the original tomographic reconstruction; here, we have $a = 7.5\,\mu$m. The scale factor M is a magnification constant: it is an integer that gives the ratio of a to the shortest distance between grid points on the cubic lattice used for the simulation of flow. We discuss the significance of M along with some other technical issues below.

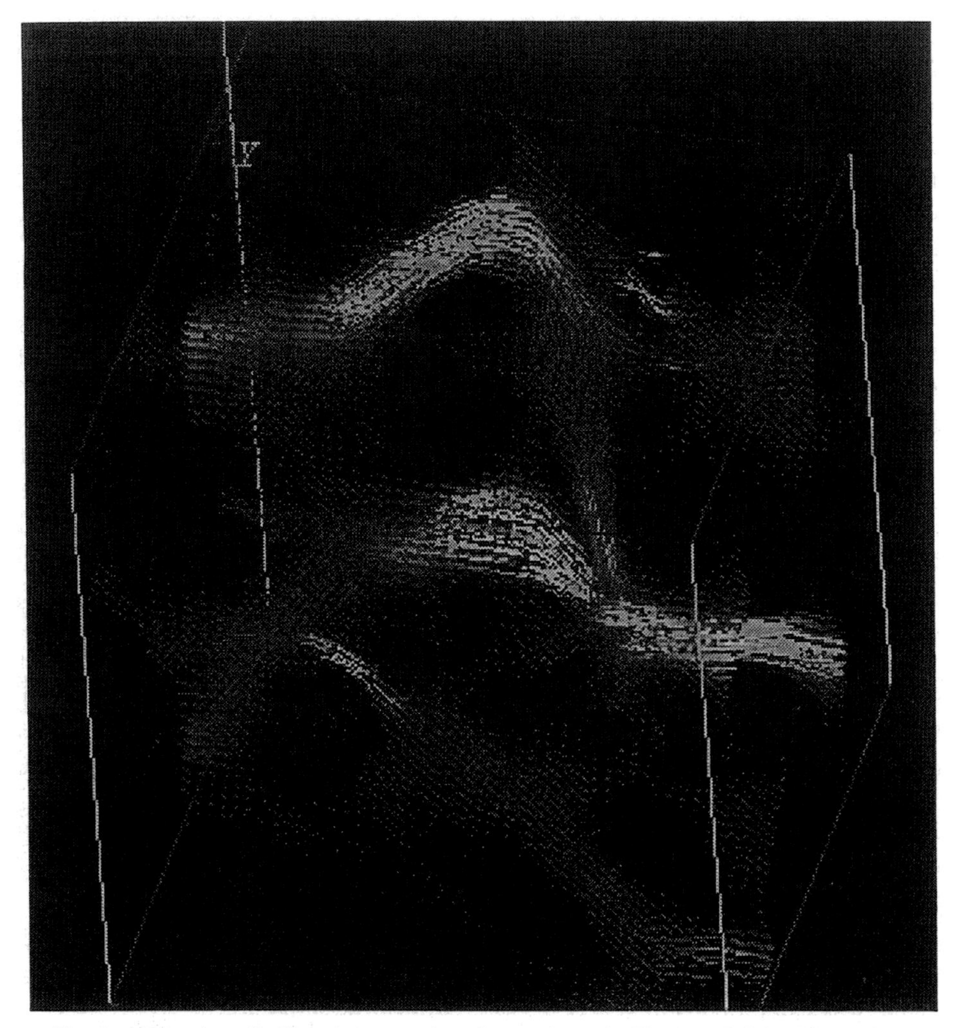

Fig. 2. Flow through the reconstructed sandstone shown in Figure 1. Bright shades of gray indicate fast flow speeds.

4.2. GRID RESOLUTION

It has long been recognized (Rothman, 1988) that applications of lattice-gas methods to flow through porous media must carefully evaluate the sensitivity of results to the resolution of the underlying lattice. Early explanations of the problem emphasized that the characteristic linear size of any region in the void space should be at least of the order of the mean free path to avoid non-hydrodynamic, 'slip-flow', or 'Knudsen flow' effects. It is now known, however, that the bounce-back boundary condition typically employed at solid-fluid boundaries does not necessarily create zero flow at the solid wall (Cornubert *et al.*, 1991; Ginzbourg and Adler, 1994). For certain choices of model parameters, and with certain wall orientations, the ideal location of the zero, halfway between the wall site and the nearest interior site, can be achieved exactly for known flows (e.g., Poiseuille or Couette flows). However, this result does not hold in general, and small deviations from the ideal zero point are possible. This problem is for practical purposes inconsequential for flows through tubes of large radii; however, if the radius of a tube is small, for example just one or two lattice units, then the effect on the resulting permeability can be substantial due to the quadratic dependence of permeability on radius.

Because we desire to simulate flow through as large a three-dimensional medium as possible, we performed an empirical study on a small piece of the rock to determine how permeability varies as a function of grid resolution – i.e., the magnification parameter M–and model parameters. For both lattice-gas and lattice-Boltzmann methods the precise location of a boundary is determined by the eigenvalues of the linearized collision operator (Cornubert *et al.*, 1991; Ginzbourg and Adler, 1994). With the relaxation-Boltzmann method, we have control over only one parameter (and thus one eigenvalue), ω. Since the viscosity, a physical variable, is determined by ω, we parameterize our study in terms of the viscosity and M. To highlight any deleterious effects, we choose a small piece of the rock that contains a narrow throat. The original linear size of the sample is 48 voxels. Several simulations were performed with $1 \leq M \leq 4$ and $0.056 \leq \nu \leq 0.5$.

The results of our tests are summarized in Figure 3. One finds, for the range of parameters studied, an approximately linear dependence of permeability on viscosity. Moreover, one finds that the viscosity dependence rapidly becomes weak for $M > 1$. The viscosity dependence is a consequence of the predicted shift of solid-fluid boundaries (Ginzbourg and Adler, 1994). The fact that the permeability varies strongly with viscosity for $M = 1$ indicates that the permeability is dominated by flow through a very narrow tube or throat, a likely possibility in such a small sample. Such a conclusion is bolstered by the results for $M > 1$; in this case the viscosity dependence is progressively weaker since perturbations to the true linear dimension of any void space scale like $1/M$.

A number of practical conclusions can be inferred from this study. First, and obviously, as the magnification parameter increases, so does the accuracy of the calculated permeability. Second, the sensitivity of the calculated permeability to magnification decreases as the viscosity decreases. Because setting $M > 1$ increases the number of grid points by a factor of M^3, below we choose to set ν small ($\nu = 0.056$) while maintaining $M = 1$. Hopefully, any errors due to the bounce-back boundary condition would probably decrease as the system size increases, since larger-scale flows should depend less on flow through very narrow tubes or throats. More precise guidelines await further, more careful, study of specific geometries (e.g., a pipe).

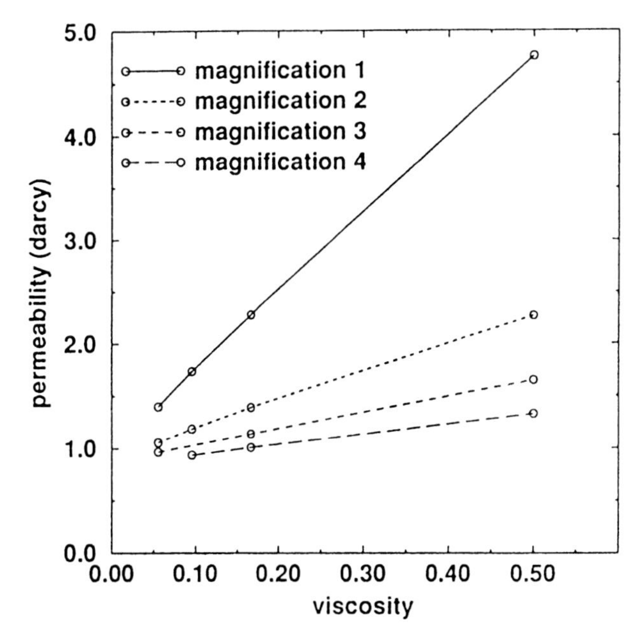

Fig. 3. Calculated permeability as a function of viscosity of a sample measuring originally 48 voxels on a side, for different magnifications of the computational grid. Magnification $M = 1$ is the unmagnified sample in which each voxel has linear dimension $a = 7.5\,\mu$m; other magnifications correspond to grid resolutions of a/M.

4.3. DEPENDENCE OF PERMEABILITY ON SAMPLE SIZE

Our largest cubic sample of Fontainebleau sandstone measures 224 voxels, or 0.17 cm, on a side. Given that we have access to the detailed pore space of the entire sample, it is interesting to ask how the permeability of small blocks of the sample compare to the permeability of larger blocks composed of these small blocks. Also, we would like to know how the permeability of the largest block compares to the experimentally measured permeability of a sample that is approximately one order of magnitude larger, and how the results from the lattice-Boltzmann method compare to what can be obtained by a finite-difference simulation of flow.

The permeability of the entire cube was measured in each orthogonal direction by the lattice-Boltzmann method. The results are displayed in Table I, where they are compared to both an earlier calculation of permeability by a finite-difference method (on a slightly larger sample) and a laboratory measurement of permeability from a core of about 4 cm in length and 2 cm in diameter (Schwartz *et al.*, 1994) located near the much smaller sample used in the tomography. Although the various numerical and experimental permeabilities scatter from 1.0 to 1.5 darcys, the scatter among the lattice-Boltzmann, finite-difference, and experimental permeabilities is well within the variability expected due to samples of differing sizes. The scatter

TABLE I. The three orthogonal permeabilities, k_x, k_y, and k_z, computed by the lattice-Boltzmann method on a sample of 224 voxels or 0.17 cm on a side. These estimates of permeability are compared to k_{fd}, the results from an earlier finite-difference /line-break calculation in the z-direction on a sample of size $288 \times 288 \times 224$ voxels (Schwartz *et al.*, 1994), and k_{exp}, the permeability obtained by laboratory experiment on a core approximately one order of magnitude larger in size (Schwartz *et al.*, 1994).

k_x	k_y	k_z	k_{fd}	k_{exp}
1.5	1.13	1.18	1.0	1.3

is also similar to the variation among calculated and laboratory measurements of permeability in a recent related study (Spanne *et al.*, 1994).

To study the permeability variation within the cubic sample of size 224, we divide the sample into 8 'mini'-blocks of linear size 112, and each of these blocks into 8 'micro'-blocks of linear size 56. A typical micro-block looks like Figure 1, only about 12% smaller in linear size.

We computed the permeability in the micro-blocks and mini-blocks in only the z-direction. Figure 4 displays that the permeability of the 64 micro-blocks varies over almost 3 orders of magnitude, with as much as an order-of-magnitude variation for the same porosity. The scatter for the mini-blocks is much smaller. We note that the overall scatter roughly follows the relation $k \propto \phi^3$.

Computation of effective or average permeabilities from constituent mini-blocks and micro-blocks provides some insight. Table II gives the arithmetic mean and harmonic mean of the permeabilities of each size studied. The arithmetic mean is defined by

$$\bar{k}_{\text{ar}} = \frac{1}{N} \sum_i k^{(i)}, \tag{6}$$

where $k^{(i)}$ is the permeability of the ith constituent block, $i = 1, \ldots, N$. The harmonic mean is

$$\bar{k}_{\text{har}} = N \left(\sum_i \frac{1}{k^{(i)}} \right)^{-1}. \tag{7}$$

One expects (Matheron, 1967)

$$\bar{k}_{\text{har}} \leqslant k \leqslant \bar{k}_{\text{ar}}. \tag{8}$$

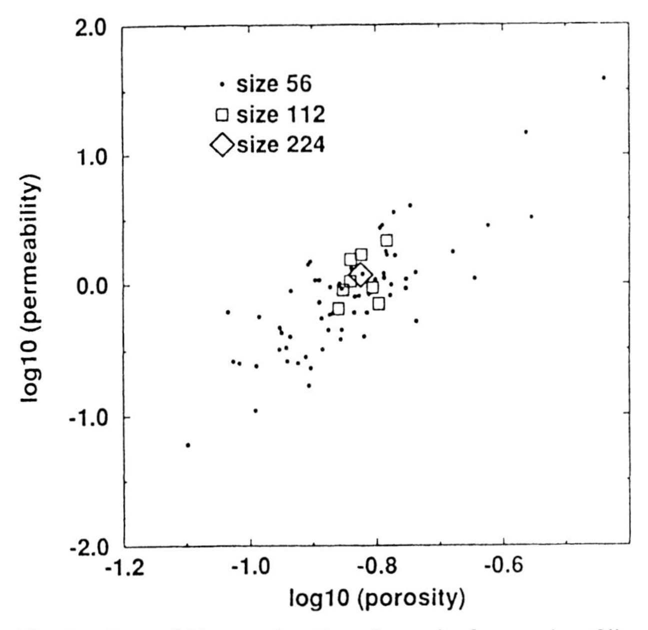

Fig. 4. Permeability as a function of porosity for samples of linear dimension 56, 112, and 224 voxels on a side, denoted by dots, squares, and a diamond, respectively.

TABLE II. Arithmetic mean $\bar{k}_{\mathrm{ar}}$ and harmonic mean $\bar{k}_{\mathrm{har}}$ for 64 samples of linear size 56 and 8 samples of linear size 112, compared to the permeability calculated for the entire cubic sample of size 224. $\delta k/\bar{k}_{\mathrm{ar}}$ is the standard deviation about $\bar{k}_{\mathrm{ar}}$, normalized to $\bar{k}_{\mathrm{ar}}$.

Size	$\bar{k}_{\mathrm{ar}}$	$\bar{k}_{\mathrm{har}}$	$\delta k/\bar{k}_{\mathrm{ar}}$
56	1.80	0.50	2.77
112	1.21	1.03	0.41
224	1.18	1.18	–

This upper and lower bound is indeed found for means computed from all 8 mini-blocks and likewise for all 64 micro-blocks.

Although the table only spans two factors of two in size, two trends in the table are worth noting. First, one sees that the arithmetic mean decreases with increasing size, while the harmonic mean increases. Second, the standard deviation decreases with size. Both observations indicate that the porous medium has a random structure with little order. In other words, as the sample sizes increase, the range of permeability variations decreases, thus causing the arithmetic and harmonic means to converge. Similar results have been found in experimental studies performed with sample sizes approximately one order of magnitude larger than we have used here (Giordano *et al.*, 1985; Henriette *et al.*, 1989).

That said, it must nonetheless be noted that the arithmetic average of the calculated permeability of each group of eight micro-blocks is not always greater than the calculated permeability of the associated mini-block. This failure to satisfy the inequality (8) is probably a consequence of jacketing the micro-blocks for the flow simulation. For small samples, the jacket can severely limit flows in directions other than the direction in which the fluid is forced, thus causing the calculated permeability to be lower than would be found in the ideal case of periodic boundary conditions.

5. Two-Phase Flow

To date, lattice-gas and lattice-Boltzmann methods are the only numerical methods that have been employed to study two-phase Navier–Stokes flow in disordered micromodels of porous media (Rothman, 1990; Gunstensen and Rothman, 1993; Soll *et al.*, 1994). Two of these studies (Rothman, 1990; Gunstensen and Rothman, 1993) were devoted to an investigation of the so-called relative permeability equations; i.e., the multiphase extension of Darcy's law. Key results included explicit demonstrations of nonlinear, non-Darcy flows in addition to numerical verification of Onsager's reciprocity for flows in the linear regime. Simulations of multiphase flow through 3D digital images of real media are not yet at such a well-developed stage, though some early results have been reported (Soll *et al.*, 1994). Below, we report preliminary qualitative results for simulations of drainage and imbibition through the piece of Fontainebleau sandstone shown in Figure 1.

In the case of drainage, the rock is initially filled with the wetting fluid and the nonwetting fluid is forced into the rock from one side. Four snapshots of the process are shown in Figure 5.

The case of imbibition is the converse: the rock is initially filled with the nonwetting fluid and the wetting fluid is forced into the rock from one side. Four snapshots of the process are shown in Figure 6, each taken at the same time as the snapshots in Figure 5. Although the wetting fluid is injected into the rock with the same force as was applied to the nonwetting fluid in the case of drainage, the wetting fluid does not advance as far because it flows along the surface of the pores, and is thus slower.

In each case the saturation of the invading fluid was calculated as function of time and space. The saturation is defined as the total mass of the invading fluid resident in each two-dimensional slice of the rock (perpendicular to the flow direction) divided by the total mass of both fluids present in the same slice. Figure 7 shows the spacetime evolution of the saturation of the nonwetting fluid during

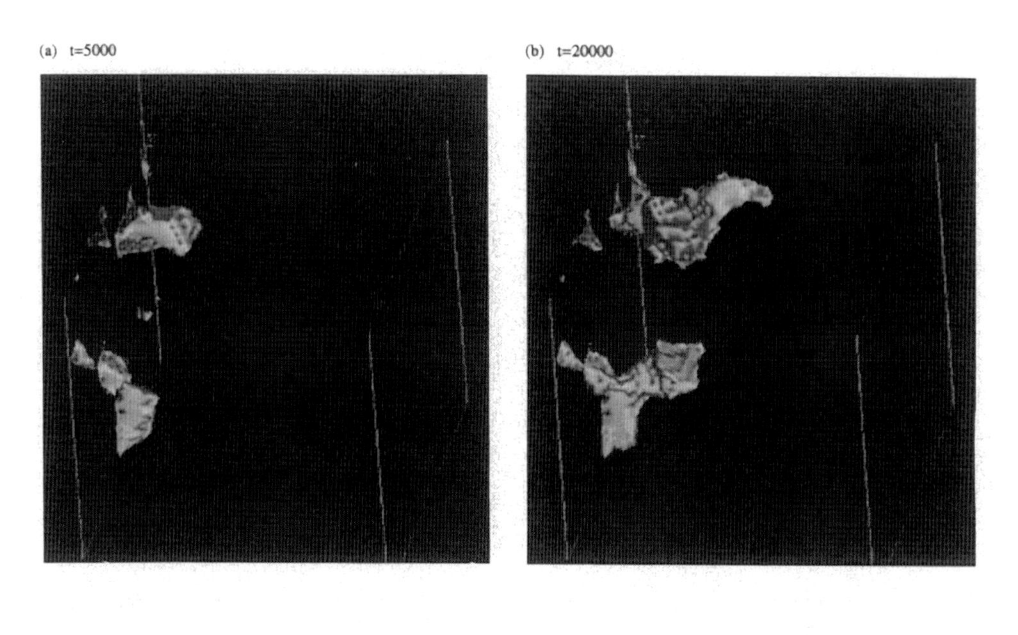

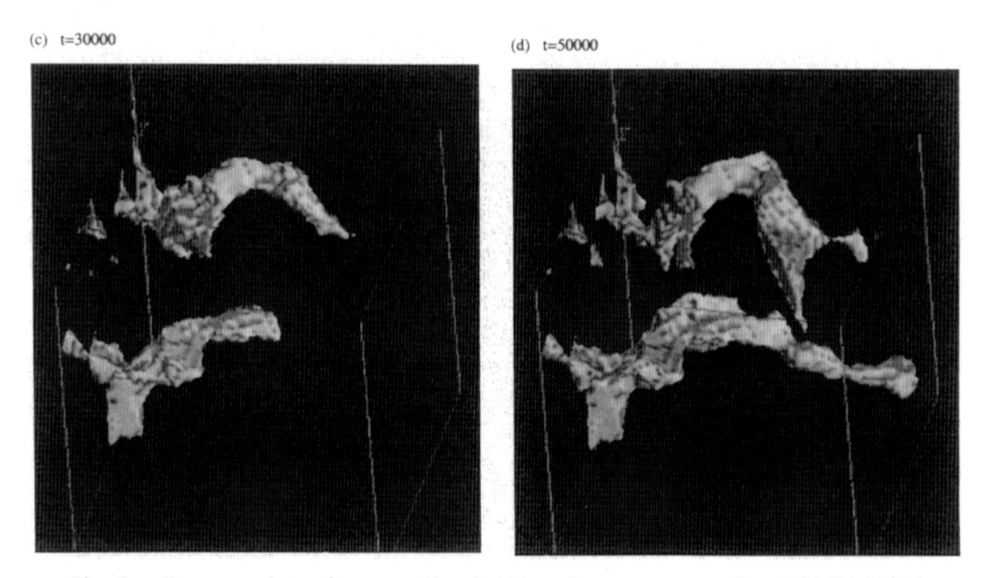

Fig. 5. Four snapshots of a nonwetting fluid invading a porous medium initially filled by a wetting fluid; the porous medium is the rock shown in Figure 1. The nonwetting fluid is injected at the left face. Parts (a), (b), (c), and (d) correspond to time steps 5000, 20 000, 30 000, and 50 000, respectively.

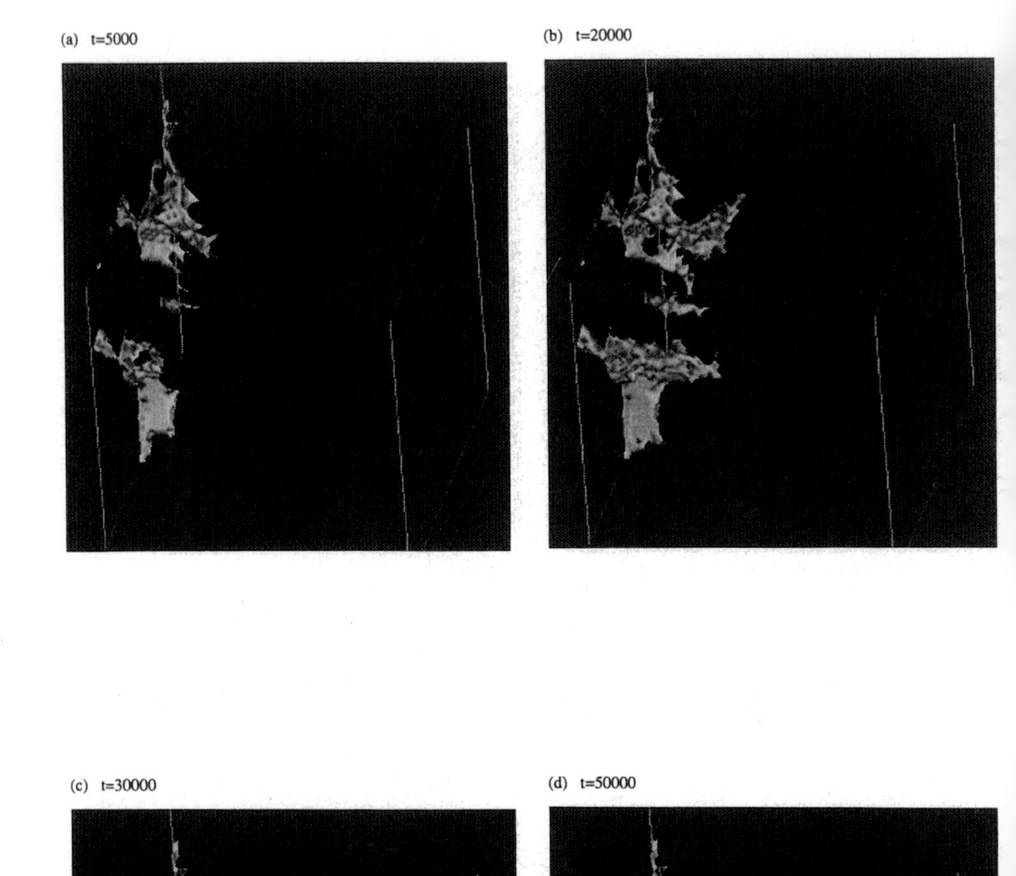

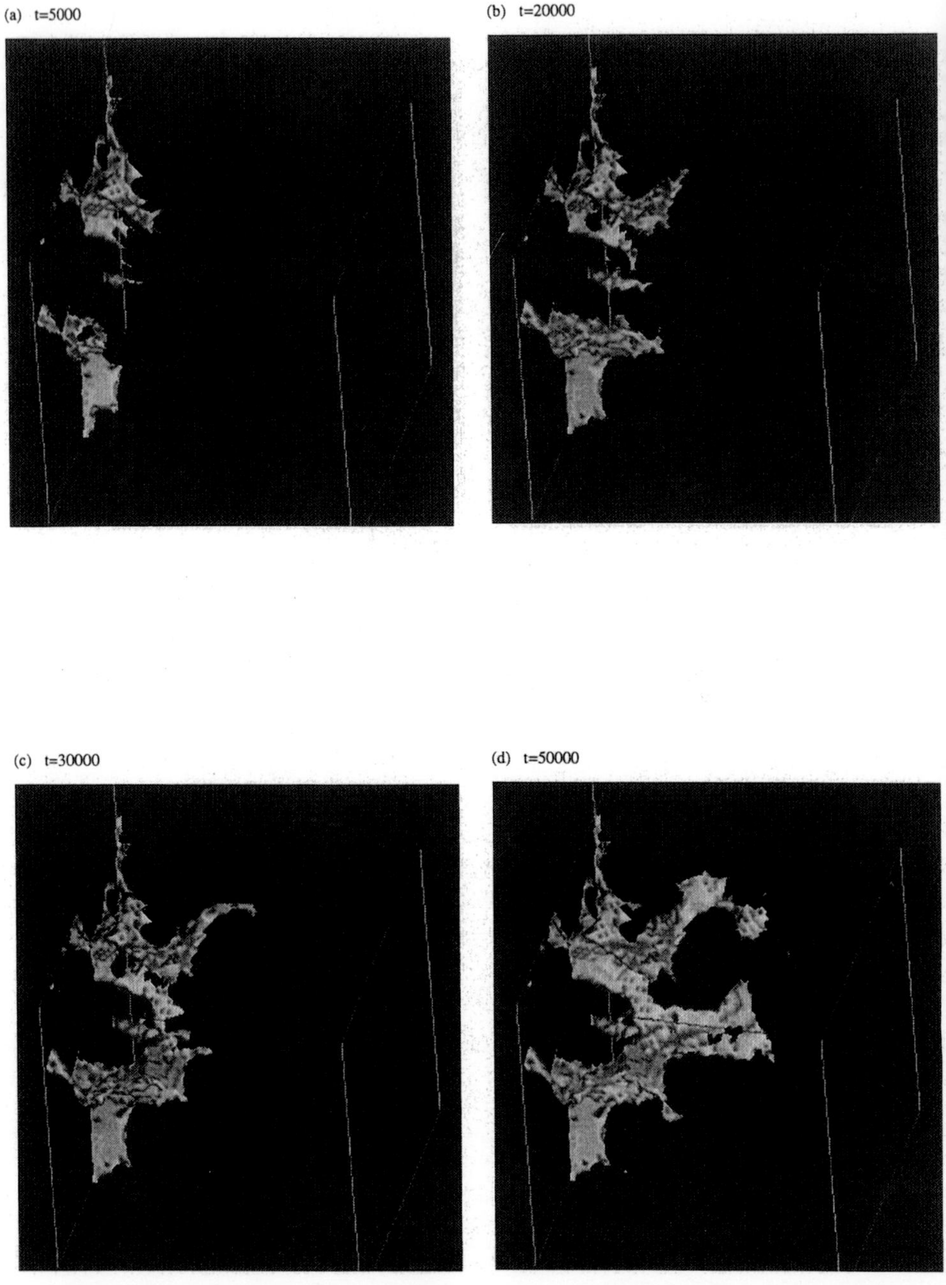

(a) t=5000
(b) t=20000
(c) t=30000
(d) t=50000

Fig. 6. Four snapshots of a wetting fluid invading a porous medium initially filled by a non-wetting fluid; the porous medium is the rock shown in Figure 1. The sequence of figures is as in Figure 5. Note that the wetting fluid progresses more slowly since it moves along the solid-fluid boundary.

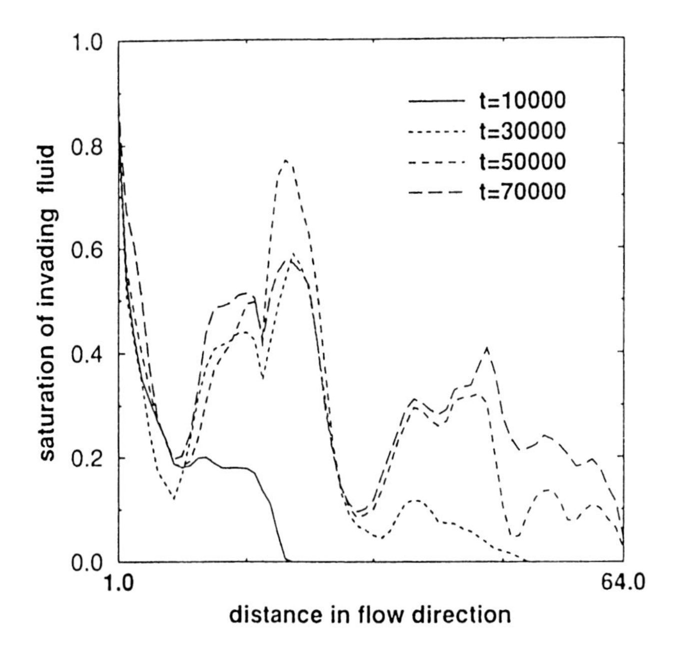

Fig. 7. Saturation of the invading (nonwetting) fluid as a function of space and time during the drainage simulation depicted in Figure 5.

drainage, and Figure 8 shows the saturation of the wetting fluid during imbibition. Although neither case reached steady state, qualitative distinctions are evident from the two plots. In the case of drainage, the maximum saturation in the interior occurs in front of and behind the large grain in the center of the rock. In other words, the nonwetting fluid has filled the biggest channels, as one expects. In the case of imbibition, however, the wetting fluid fills the rock relatively uniformly. Indeed, the plot of saturation shows the growth of a sharp front marking the advancement of the wetting fluid.

6. Discussion and Conclusion

In this paper, we have provided an overview of the 'state-of-the-art' of lattice-gas and lattice-Boltzmann simulations of flow through porous media. In particular, we have concentrated on studies of flow through a 3D tomographic reconstruction of Fontainebleau sandstone at the scale of about one-half to two millimeters in linear dimension.

The most important result that we have reported relates to our permeability calculations. Specifically, we find that our lattice-Boltzmann simulations find approximately the same permeability as related finite-difference calculations and laboratory measurements. The scatter of all these measurements, which vary from

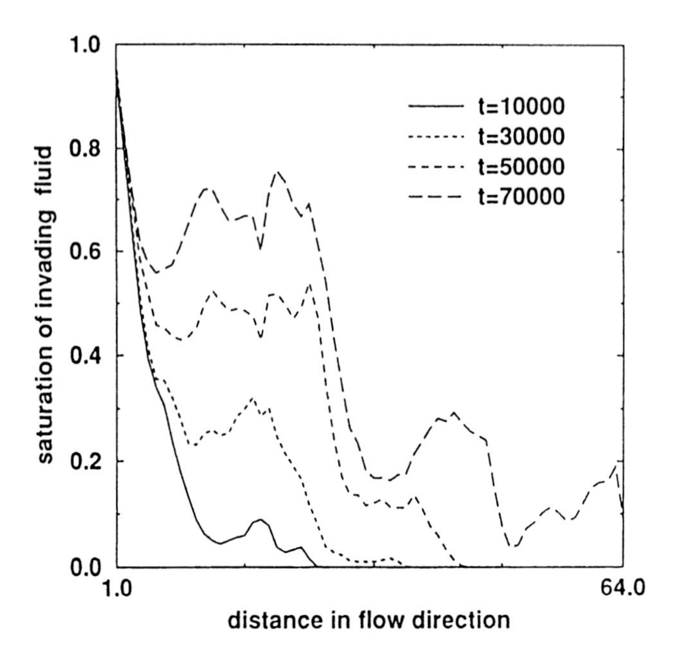

Fig. 8. Saturation of the invading (wetting) fluid as a function of space and time during the imbibition simulation depicted in Figure 6.

about 1.0 to 1.5 darcys, is thought to be due to a combination of size-dependent permeabilities and numerical inaccuracies. A similar scatter of calculated and laboratory measurements was recently reported by Spanne *et al.* (1994).

We also reported a study of the statistical variation of permeability within the same sample. We find that the permeability variations are greatest for the smallest samples, an observation which is consistent with a pore space that is randomly constructed. At larger scales, however, physical processes related to the formation of sedimentary rocks may induce spatial correlations (Thompson *et al.*, 1987). One possible use of microscopic flow calculations in the future may be to detect at which scale these correlations begin to affect transport.

Lastly, we reported a qualitative study of two-phase flow, for the cases of drainage and imbibition. The results were roughly in accord with the expected behavior.

One aspect of the state-of-the-art that we emphasized here is the dependence of permeability calculations on grid resolution. Our empirical calculations make clear that a pore-space that contains very narrow channels or tubes, on the order of 1 lattice unit in radius, may produce significant errors in calculated permeabilities if the bulk flow is dominated by narrow passageways. Although such errors are difficult to precisely quantify, we feel that the accuracy of our results is in the neighborhood of 10–50%, with errors decreasing as the sample size increases. We note that a dependence of permeability on grid resolution is also expected for finite-difference solutions to the Stokes equations, if for no other reason than that the quality of numerical integration increases as space discretization becomes finer. If more accurate estimation of bulk transport properties is an important goal, the relatively efficient Boolean lattice-gas method may well be the method of choice. Such studies will be reported in the near future.

Acknowledgements

We thank Larry Schwartz for providing the microtomography data and for several discussions. Additionally we would like to thank Dominique d'Humières, Nick Martys, and Art Thompson for offering their insight about our numerical results. This work was supported in part by the sponsors of the MIT Porous Flow Project, NSF Grant 9218819-EAR, and Elf Aquitaine. The authors additionally wish to acknowledge the Advanced Computing Laboratory of Los Alamos National Laboratory, Los Alamos, NM 87545. This work was partially performed on computing resources located at this facility.

References

Adler, P.: 1992, *Porous Media: Geometry and Transports*, Butterworth/Heinemann, London.

Benzi, R., Succi, S., and Vergassola, M.: 1992, The lattice Boltzmann equation: Theory and applications, *Phys. Rep.* **222**, 145–197.

Cancelliere, A., Chang, C., Foti, E., Rothman, D., and Succi, S.: 1990, The permeability of a random medium: Comparison of simulation with theory, *Phys. Fluids A* **2**, 2085.

Chandler, R., Koplik, J., Lerman, K., and Willemsen, J.: 1982, Capillary displacement and percolation in porous media, *J. Fluid Mech.* **119**, 249–267.

Chen, H., Chen, S., and Matthaeus, W. H.: 1992, Recovery of the Navier–Stokes equations using a lattice gas Boltzmann method, *Phys. Rev. A* **45**, R5339–5342.

Chen, S., Diemer, K., Doolen, G., Eggert, K., Fu, C., Gutman, S., and Travis, B. J.: 1991, Lattice gas automata for flow through porous media, *Physica D* **47**, 72–84.

Cornubert, R., d'Humiéres, D., and Levermore, D.: 1991, A Knudsen layer theory for lattice gases, *Physica D* **47**, 241.

Cushman, J. H. (ed): 1990, *Dynamics of Fluids in Heirarchical Porous Media*, Academic Press, San Diego.

Dias, M. and Payatakes, A.: 1986a, Network models for two-phase flow in porous media, Part 1. Immiscible microdisplacement of non-wetting fluids, *J. Fluid Mech.* **164**, 305–336.

Dias, M. and Payatakes, A.: 1986b, Network models for two-phase flow in porous media, Part 2. Motion of oil ganglia, *J. Fluid Mech.* **164**, 337–358.

Flannery, B. P., Deckman, H. W., Roberge, W. G., and D'Amico, K. L.: 1987, Three-dimensional X-ray microtomography, *Science* **237**, 1439–1444.

Frisch, U., d'Humières, D., Hasslacher, B., Lallemand, P., Pomeau, Y., and Rivet, J.-P.: 1987, Lattice gas hydrodynamics in two and three dimensions, *Complex Systems* **1**, 648.

Frisch, U., Hasslacher, B., and Pomeau, Y.: 1986, Lattice-gas automata for the Navier–Stokes equations, *Phys. Rev. Lett.* **56**, 1505–1508.

Ginzbourg, I. and Adler, P. M.: 1994, Boundary flow condition analysis for the three-dimensional lattice-Boltzmann model, *J. Phys. II France* **4**, 191–214.

Giordano, R. M., Salter, S. J., and Mohanty, K.: 1985, SPE 14365: The effects of permeability variations on flow in porous media, in *Proc. 60th Ann. Tech. Conf. and Exhibition Soc. Petroleum Engrs.*, Las Vegas, Nevada, September 22–25, 1985.

Gunstensen, A. K. and Rothman, D. H.: 1992, Microscopic modeling of immiscible fluids in three dimensions by a lattice-Boltzmann method, *Europhys. Lett.* **18**(2), 157–161.

Gunstensen, A. K. and Rothman, D. H.: 1993, Lattice-Boltzmann studies of two-phase flow through porous media, *J. Geophys. Res.* **98**, 6431–6441.

Gunstensen, A. K., Rothman, D. H., Zaleski, S., and Zanetti, G.: 1991, A lattice-Boltzmann model of immiscible fluids, *Phys. Rev. A* **43**, 4320–4327.

Henriette, A., Jacquin, C. G., and Adler, P. M.: 1989, The effective permeability of heterogeneous porous media, *Physico-Chem. Hydrodyn.* **11**, 63–80.

Kohring, G.: 1991, Calculation of the permeability of porous media using hydrodynamic cellular automata, *J. Stat. Phys.* **63**, 411-418.

Koplik, J. and Lasseter, T.: 1985, One- and two-phase flow in network models of porous media, *Chem. Eng. Comm.* **26**, 285–295.

Martys, N. and Garboczi, E. J.: 1992, Length scales relating the fluid permeability and electrical conductivity in random two-dimensional model porous media, *Phys. Rev. B* **46**, 6080.

Martys, N., Torquato, S., and Bentz, D.: 1994, Universal scaling fluid permeability for sphere packings, *Phys. Rev. E* **50**, 403.

Matheron, G.: 1967, *Eléments pour une théorie des milieux poreux*, Masson, Paris.

Qian, Y., D'Humiéres, D., and Lallemand, P.: 1992, Lattice BGK models for Navier–Stokes equation, *Europhys. Lett.* **17**(6), 479–484.

Rothman, D. H.: 1988, Cellular-automation fluids: A model for flow in porous media, *Geophysics* **53**, 509–518.

Rothman, D. H.: 1990, Macroscopic laws for immiscible two-phase flow in porous media: Results from numerical experiments, *J. Geophys. Res.* **95**, 8663.

Rothman, D. H. and Zaleski, S.: 1994, Lattice-gas models of phase separation: interfaces, phase transitions, and multiphase flow, *Rev. Modern Phys.* **66**, 1417–1479.

Schwartz, L. M., Auzerais, F., Dunsmuir, J., Martys, N., Bentz, D. P., and Torquato, S.: 1994, Transport and diffusion in three-dimensional composite media, *Physica A* **207**, 28–36.

Schwartz, L. M., Martys, N., Bentz, D. P., Garboczi, E. J., and Torquato, S.: 1993, Cross-property relations and permeability estimation in model porous media, *Phys. Rev. E* **48**, 4584.

Soll, W., Chen, S., Eggert, K., Grunau, D., and Janecky, D.: 1994, Application of the lattice-Boltzmann/lattice gas technique to multi-fluid flow in porous media, in A. Peters (ed), *Computational Methods in Water Resources X*, Kluwer Acad. Publ. Dordrecht, pp. 991–999.

Spanne, P., Thovert, J. F., Jacquin, C. J., Lindquist, W. B., Jones, K. W., and Adler, P. M.: 1994, Synchrotron computed microttomography of porous media: topology and transports, *Phys. Rev. Lett.* **73**, 2001–2004.

Thompson, A. H., Katz, A. J., and Krohn, C. E.: 1987, The microgeometry and transport properties of sedimentary rock, *Adv. Physics* **36**, 625–694.

Transport in Porous Media **20:** 21–35, 1995.

Simulation of Capillary-Dominated Displacements in Microtomographic Images of Reservoir Rocks

R. D. HAZLETT
Mobil Exploration and Producing Technical Center, 13777 Midway Rd., Dallas, TX 75244, U.S.A.

(Received: May 1994)

Abstract. Displacement simulation in realistic pore networks, such as those derived from X-ray microtomography, is presented for the regime where capillarity controls fluid motions and spatial distributions. Complex displacement sequences involving both imbibition and drainage are constructed to extract wettability indices. The percolation properties of predicted equilibrium phase distributions are analyzed. Equilibrium fluid distributions are used to model transport properties for each phase.

Key words: capillarity, networks, microtomography, percolation.

1. Introduction

Network models have long been used to study displacement mechanisms in idealized pore systems [1–5]. Seldom do such models claim to be representative of reservoir rocks. Imaging techniques now offer the ability to examine actual pore networks in three dimensions at micrometer resolution [6–8]. One such example is provided in Figure 1. Depicted is a 2 mm^3 subset of a 1 μm^2 Berea sandstone X-ray-computed tomography data set which was acquired at 10 μm resolution [7]. Transport property computations on such arbitrary networks present formidable challenges. Such work has progressed for single-phase properties in the absence of measured pore networks with stochastically generated media [9, 10] and most recently with microtomography data [11].

In order to extend predictive capabilities to multiphase systems, an algorithm to conduct invasion percolation simulations on such networks with arbitrary pore shapes and realistic pore interconnections was constructed. Displacements can be computed as the result of either increasing or decreasing the capillary pressure of the system. Mixed wettability systems are considered by dividing the overall network into interacting water-wet and oil-wet sub-networks. Complex displacement sequences such as those in the combined USBM–Amott test [12] can be performed, allowing computation of wettability indices for mixed-wet systems of varying water-wet volume fraction. Phase distributions can be examined for continuity in each phase as a function of saturation and saturation history. Spatial distributions of fluid phases can also be used to compute steady state transport property estimates for each phase and hysteresis in those values.

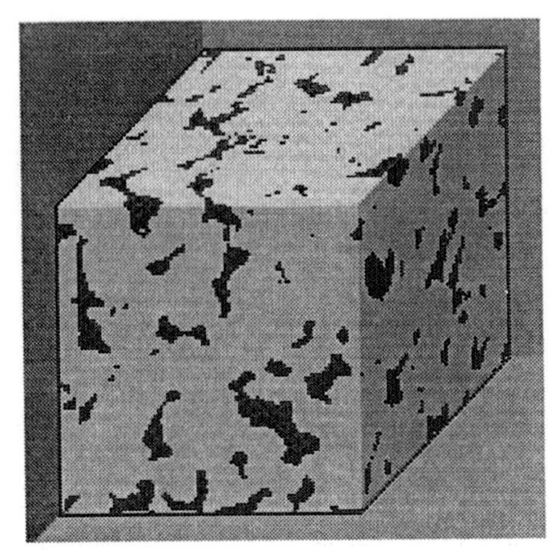

Fig. 1. X-ray microtomographic image of 2 mm^3 of Berea sandstone which was acquired at 10 μm resolution (data courtesy of Jasti [7]).

2. Conventions

Due to natural surface roughness in reservoir rocks, it has been proposed that oil reservoir wettability may often be approached as a binary property [13, 14]. That is, one phase will always perfectly wet the solid. Mixed-wet states are not disallowed in this view of porous media but are the result of perfectly wetting subdomains with different fluid preferences. These preferences can be directly associated with mineralogy distributions [15], or they could be the result of absorption or deposition of organic species over surfaces in contact with oil [16]. Unless specified otherwise, wettability will herein be considered a binary property with contact angles being either 0° or 180°.

Capillary dominated displacements are controlled by stable, equilibrium configurations of fluid/fluid interfaces in a pore system. At any given capillary pressure, the menisci separating bulk fluid phases will all have the same mean curvature if the system is truly an equilibrium state. Subject to the wettability constraint, in this work, fluid-fluid boundaries are represented by the envelope of the union of spherical cap segments with a radius greater than a common threshold. The threshold radius is related to a capillary pressure through the Laplace equation. Fluid advancement, in a quasi-static sense, is then governed by propagation of spherical menisci or the envelope of assemblages of spherical caps through the pore network.

Fluid displacements are subject to connectedness constraints in the absence of substantial mutual solubility between bulk fluid phases. Nonwetting fluid displacement can occur only along hydraulically connected pathways of the same phase. Otherwise, fluid motion does not take place. Should connectedness be lost, residual phase stranding occurs. For our purposes, the wetting phase is assumed to always

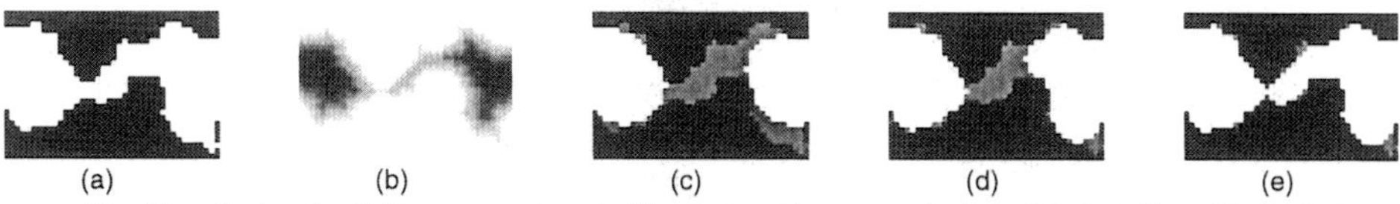

Fig. 2. A simple, 2-D pore system to illustrate volume rendering. (a) the 27×40 pixel pore system, (b) the inscribed sphere radius map, (c) accessibility-independent nonwetting phase distribution at $R_c = 6.5$, (d) nonwetting phase distribution at $R_c = 4.5$, (e) nonwetting phase distribution at $R_c = 2.5$.

maintain connectedness. This is reasonable since microporosity and surface roughness are not adequately represented in images of typically 5–10-μm resolution. Sub-micron features are known to provide conduits for fluid migration and are important when considering equilibrium conditions and transport properties [5, 17].

3. Methodology

In general, it is the nonwetting phase which is tracked and the wetting phase is identified by comparison with the full porosity image. Operations, unless specified otherwise, are on binary fields with a flag indicating nonwetting phase occupancy. All computations were performed using Fortran 77 on an IBM R6000.

Given a network which has been reduced to a map of pore versus rock, an inscribed sphere map is constructed as follows. At each voxel designated as pore space, a number is assigned representing the radius of the largest sphere which could be placed at that location without intersecting rock. The inscribed sphere mask provides the framework for developing equilibrium saturation distributions. An example of such an inscribed sphere mask for a two-dimensional model pore system is shown in Figure 2. This information can be used to generate a phase map for a given capillary pressure.

Phase maps can be produced by rendering a subset of the inscribed spheres whose radii were determined in the previous step. The union of spheres with radii greater than or equal to a cutoff radius would represent a distribution of nonwetting phase during primary drainage if all pore space were accessible. One can compute such accessibility-independent phase distributions for the range of radii present in the inscribed sphere mask. Such a rendering with a radius of half of a voxel would represent the entire porosity image. Each cutoff radius would correspond to a different level of capillary pressure according to the Laplace equation. Accessibility-independent phase mappings for three different cutoff radii have also been provided in Figure 2. The smaller the cutoff radius, the more the nonwetting phase conforms to the inherent surface roughness.

However, fluid must either enter or exit through the exterior surfaces. Again, it is assumed that the wetting phase may always be drained should occupancy by nonwetting phase be favored. The actual nonwetting phase distribution is that

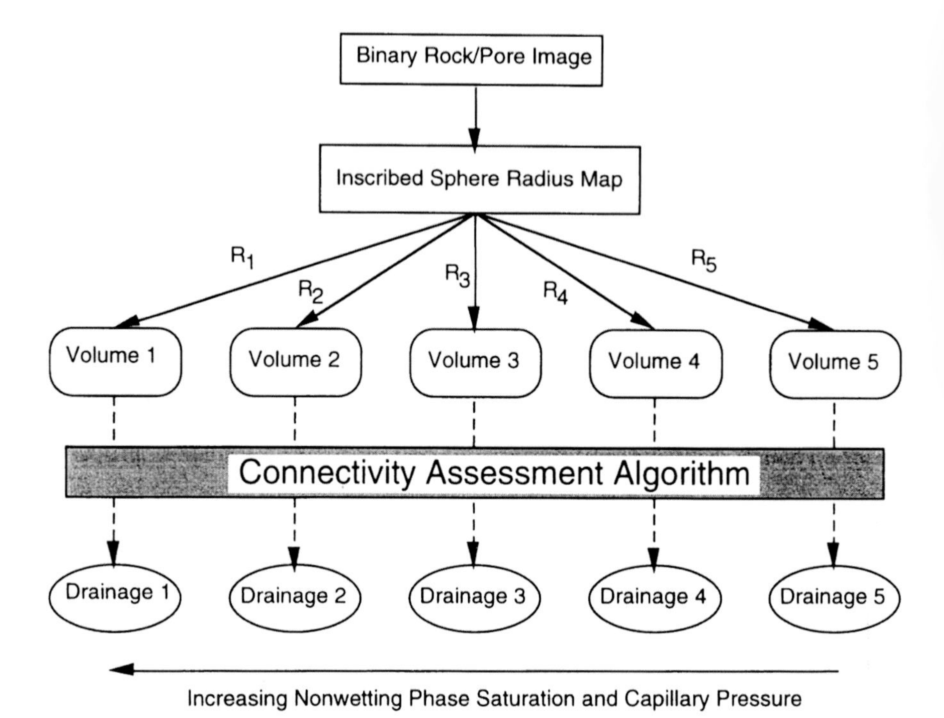

Fig. 3. Flow diagram for primary drainage computations.

subset of the sphere union mask which is connected to the face(s) where injection or production can take place. For the simple, 2-D example in Figure 2, the role of accessibility is rather obvious. Should nonwetting phase only be allowed to enter from either the left or the right, only a subset of the volume rendered can actually be filled. A program which runs a connectivity check on each voxel identified as nonwetting phase in a rendered sphere volume and excludes those elements without a pathway to user defined bounding surfaces was constructed. Resulting phase distributions represent predicted equilibrium configurations found upon primary drainage. This entire process is illustrated in Figure 3 for an image with only five different radii in the inscribed sphere mask.

In imbibition simulation (decreasing capillary pressure), when considering a new equilibrium state, all pore space available for the wetting phase in the sphere map with the appropriate radius of curvature is accessible to the wetting phase. However, the nonwetting phase must have a continuous nonwetting phase pathway to the exterior, or phase trapping results. A new imbibition state is found by forming the union between nonwetting phase present which is immobile and that fraction of existing nonwetting phase which is already in the preferred accessibility-independent configuration at the target capillary pressure. The nonwetting phase which can be stranded at each incremental decrease in capillary pressure is found by identifying that fraction of existing nonwetting phase which is not in the appropriate

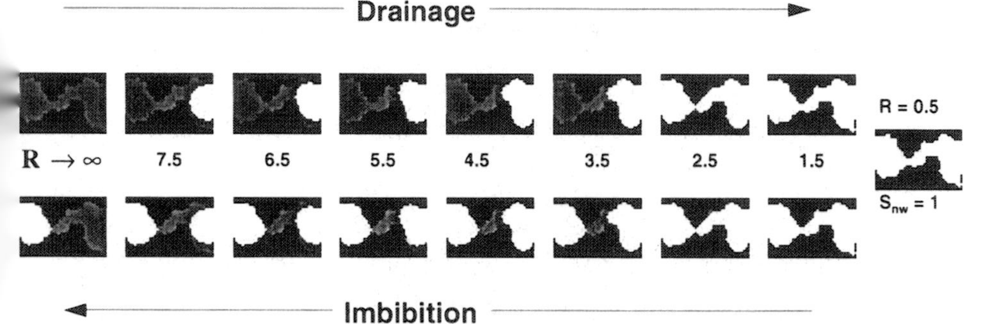

Fig. 4. Drainage and imbibition simulation in the simple, 2-D model pore system of Fig. 2, where nonwetting phase can only enter or exit from the right. Nonwetting phase is white, and wetting phase is gray. Pore entry radii are indicated.

primary drainage image, since the nonwetting phase in that image is connected to the sample exterior. As an example, the set of operations required to construct imbibition images following complete intrusion of nonwetting phase as diagrammed in Figure 3 is provided below.

$$\begin{aligned}
\text{new state} &= (\text{immobile nonwetting phase}) \cup \\
&\quad (\text{nonwetting phase in the preferred configuration}) \\
(\text{Imbibition } 2) &= \emptyset \cup ((\text{Drainage } 1) \cap (\text{Volume } 2)) \\
(\text{Imbibition } 3) &= ((\text{Imbibition } 2) \cap (\overline{\text{Drainage } 2})) \cup \\
&\quad ((\text{Imbibition } 2) \cap (\text{Volume } 3)) \\
(\text{Imbibition } 4) &= ((\text{Imbibition } 3) \cap (\overline{\text{Drainage } 3})) \cup \\
&\quad ((\text{Imbibition } 3) \cap (\text{Volume } 4)) \\
(\text{Imbibition } 5) &= ((\text{Imbibition } 4) \cap (\overline{\text{Drainage } 4})) \cup \\
&\quad ((\text{Imbibition } 4) \cap (\text{Volume } 5)) \\
(\text{Imbibition } 6) &= (\text{Imbibition } 5) \cap (\overline{\text{Drainage } 5}).
\end{aligned}$$

In this set representation, the complement of a nonwetting phase distribution, designated by the bar, represents both wetting phase and rock; however, the intersection operation identifies trapped nonwetting phase and excludes the rock. The accessibility-independent nonwetting phase images are designated by Volume i to be consistent with Figure 3, where i is the cutoff inscribed sphere radius. The initial state for the imbibition process is a drainage endpoint which contains no immobile nonwetting phase; therefore, the first contribution to the initial imbibition step is the null set, $\emptyset$.

For illustration purposes, drainage and imbibition simulation was carried out on the simple, 2D model pore system of Figure 2, subject to the described assumptions

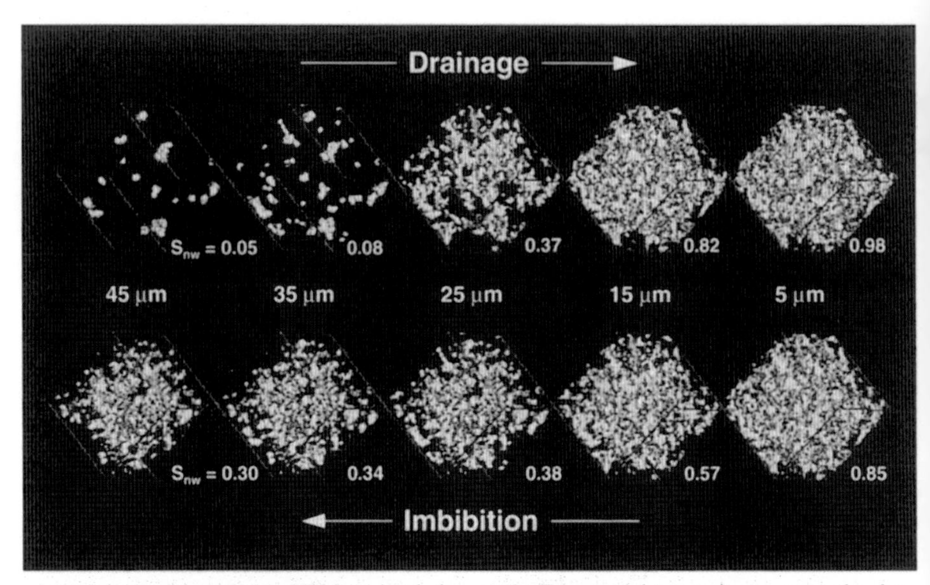

Fig. 5. Drainage and imbibition simulation on the Berea sandstone microtomography data set. Only the nonwetting phase is depicted. Radii of curvature and corresponding saturations are provided. All faces are accessible.

on phase connectivity. The phase distributions as a function of changes in pore entry radius are provided in Figure 4. In this simulation, the nonwetting phase can only enter or exit from the right. Invasion of the pore body at the open surface is allowed at a pore entry radius of 7.5 pixels, but penetration is not allowed to proceed due to the presence of a pore throat. Decreasing pore entry radius causes filling of pore roughness, as the method approximates the solid curvature locally with tangent circles of decreasing size. When the pore entry diameter equals the pore throat diameter, the downstream pore body is filled. This is the same point at which percolation is observed. At a pore entry radius of one half of a pixel, all pore space is occupied by nonwetting phase. Upon decreasing capillary pressure, the saturation pathway is retraced until the equivalent pore entry radius increases beyond a value of 2.5 pixels. At this point, phase disconnection is favored over pore emptying, and stranding occurs in the leftmost pore body. This type of stranding is analogous to the so-called ink-bottle effect [18] except that pores need not have only a single entrance or exit to be susceptible to phase trapping. In contrast, the contents of the pore connected to the exterior is allowed to extrude completely.

As a more complex example of the procedure on a realistic pore network, drainage and imbibition were simulated using the Berea network of Figure 1. Resulting nonwetting phase maps for the Berea sandstone network are provided in Figure 5. In this simulation, all six exterior faces were available for nonwetting phase communication; hence, this simulation is more representative of mercury porosimetry experiment. The radius of curvature corresponding to a given capillary pressure is provided along with the nonwetting phase saturation. The rock and wetting phase have been eliminated from the image for clarity. Although not completely apparent without rotation of the images, all nonwetting phase in the drainage simulation is connected to one of the six faces. Two percent of the porosity in this data set remained unconnected at the pore entry size corresponding to the image resolution. Prior to imbibition simulation, it was assumed that the capillary pressure was further increased to allow invasion into those pores as well. This is not a limitation but merely an assumption for this set of computations. As will be illustrated later, imbibition can commence at any equilibrium saturation distribution. For this simulation, a residual saturation of 30% is found which is reasonable for Berea sandstone.

In order to illustrate the complex set of operations in imbibition simulation, a single imbibition step has been broken down into this elemental parts. In particular, the series of intersections and unions has been performed in the Berea pore network for the nonwetting saturation change from 57 to 38%, as seen in Figure 5. The images involved in the various operations are presented in Figure 6. The top row extracts that portion of nonwetting phase which exists in the present state but is not a part of the nonwetting phase network which is connected to a designated exit face at the current capillary pressure. This nonwetting phase is disconnected or trapped. The remainder of nonwetting fluid at the current state is mobile. A fraction of the nonwetting phase is already in the accessibility-independent configuration of the new, lower capillary pressure. The middle row in Figure 6 depicts the intersection of the current state with the accessibility-independent configuration at the target capillary pressure. In the example provided, the target accessibility-independent distribution is a subset of the current distribution. This second operation does not require the isolation of the mobile fraction at the current state since all the trapped phase is retained in the final union operation anyway. The union of the two nonwetting distributions gives the new configuration at the lower capillary pressure.

It should be emphasized that the phase distributions predicted during drainage and imbibition depend upon the boundary conditions imposed and the saturation history. The boundary conditions of the simple two-pore model and the sandstone pore network illustrate two common experiments of interest. In the former, nonwetting phase transfer was permitted from only a single direction. In contrast, transport was allowed at all faces in the latter. Since capillary pressure experiments can be performed in a variety of modes, the impact of the number of accessible faces will be a subject of considerable interest for the future.

In secondary drainage, the residual nonwetting phase saturation must be considered in conjunction with the sphere mask. The union of the residual phase mapping and the sphere mask is constructed. The connectivity of the proposed nonwetting phase saturation is then assessed, again disregarding portions which cannot be traced to an injection face. The final step is to reintroduce any portion of the original residual phase saturation which would have been eliminated during the continuity assessment.

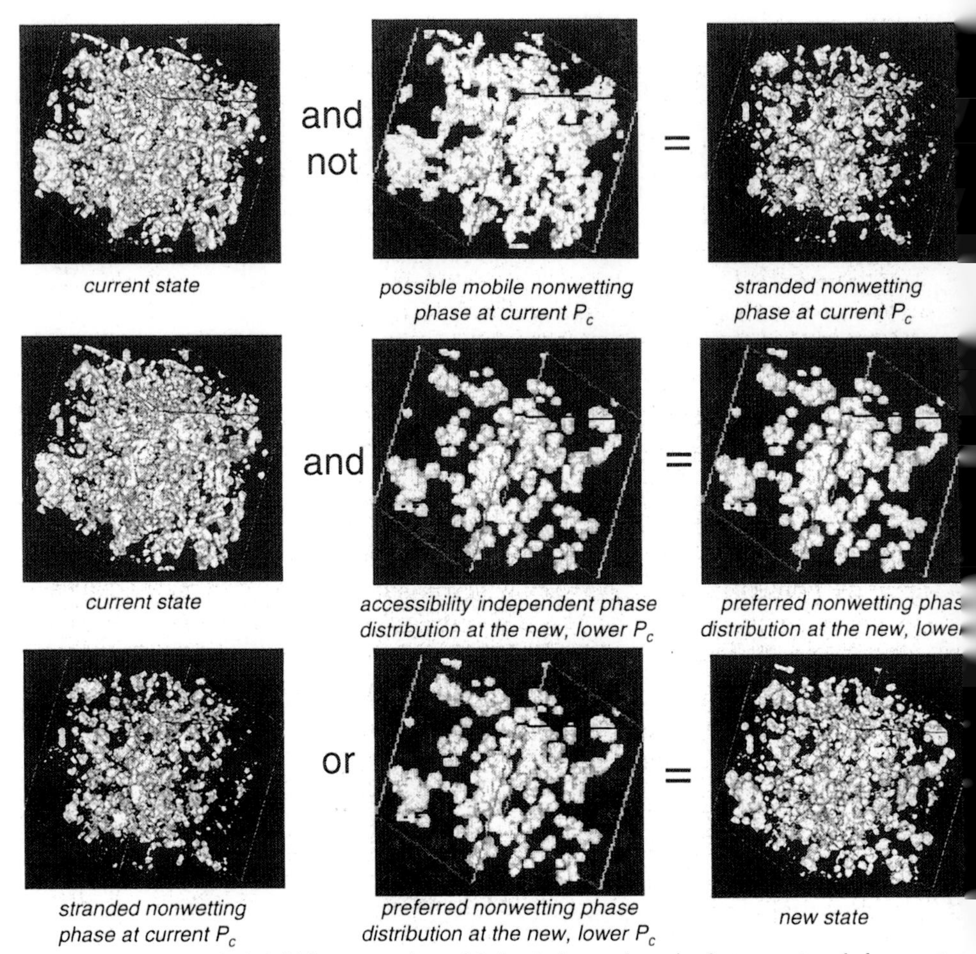

Fig. 6. Steps in imbibition network model simulation as the union between stranded nonwetting phase at the current capillary pressure and the nonwetting phase already in the preferred accessibility independent configuration of the target capillary pressure.

Mixed wettability systems can be handled as a natural extension of the procedures outlined. Herein, it is assumed that mixed-wet networks are the result of adsorption or deposition on surfaces contacted by an oil phase. Thus, a mixed-wet network is constructed using the fluid distributions from primary drainage. That fraction of the porosity network filled with oil is considered an oil-wet sub-network, while the remainder is still water-wet. Rules are similar to those in secondary drainage except one must consider the union of phase distributions on both networks, one increasing wetting phase saturation and the other decreasing wetting phase saturation. Connectivity assessments, when required, are performed on the union of the two sub-networks. The wetting phase on each sub-network is assumed to maintain connectivity via films. At this point, investigation of any series of displacements is reduced to a complex accounting problem.

4. Wettability Indices

Since drainage and imbibition experiments can be simulated, specific displacement sequences such as those in determining wettability indices can be performed. One such sequence of particular interest is the combined USBM–Amott test [12]. In this test, end-points are typically defined at reference points of ± 10 psi (68.9 kPa) capillary pressure. In this experimental suite, the porous medium is first brought to a low initial water saturation. Water is then allowed to imbibe spontaneously. This is followed by a water-drive displacement. Spontaneous oil imbibition is permitted, and an oil-drive displacement completes the sequence. The Amott indices, δ_w and δ_o, note the fractional saturation change occurring spontaneously from unforced and forced displacements, both of oil and brine.

$$\delta_w = V_{osp}/V_{ot}, \tag{1}$$

$$\delta_o = V_{wsp}/V_{wt}, \tag{2}$$

$$I = \delta_w - \delta_o. \tag{3}$$

V_{osp} and V_{wsp} are the volumes of oil and water which imbibe spontaneously, and V_{ot} and V_{wt} are the volume totals for each phase from both spontaneous imbibition and forced imbibition. The combined index, I, indicates the overall behavior. In contrast, the USBM method is a ratio of the work done during oil and brine displacements. The USBM index, W, is computed from the area under the capillary pressure curve

$$W = \log(A_+/A_-). \tag{4}$$

The areas above and below the zero capillary pressure condition are denoted by A_+ and A_-, respectively. The combined USBM–Amott test procedure allows both types of wettability indices to be computed as complementary measures of wetting character.

Combined USBM–Amott tests were simulated for two different mixed-wettability scenarios in the Berea sandstone geometry. These cases are distinguished by the initial water saturation present before subdivision of the network into oil and water-wet domains. Only those cases where the oil phase forms a percolating network can actually be classified as mixed-wettability states. In the preferred terminology, nonpercolating systems should be classified as fractional wettability states. In the first case, 64% of the network remained water-filled, and thus water-wet. The displacement series is illustrated in Figure 7. The wettability indices recovered are indicative of a weakly water-wet system. In the second case,

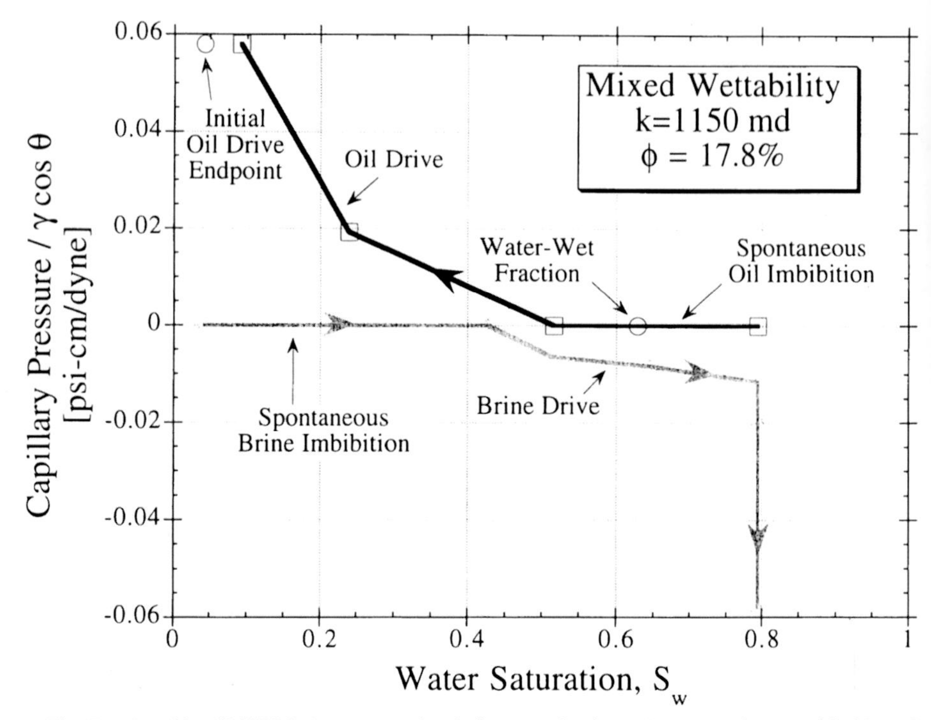

Fig. 7. Combined USBM–Amott test simulation on mixed-wet Berea sandstone with 64% of the pore system designated as water-wet. USBM index, $W = 0.13$; Amott indices: $\delta_w = 0.51$, $\delta_o = 0.40$, $I = 0.11$.

only 18% of the total porosity network by volume was water-wet. The constructed capillary pressure curve is provided as Figure 8. Wettability indices are characteristic of a strongly oil-wet system. Thus, the results of combined USBM–Amott simulation using the capillary dominated displacement model are very reasonable. Results are consistent with observed trends in wettability indices with respect to initial brine saturation in Berea sandstone using crude oils which are known to alter wettability [19].

5. Percolation Properties

The predicted equilibrium fluid distributions were tested for continuity over the extent of the pore network. A phase with a continuous pathway spanning the system is said to percolate. Percolation theory was developed upon the premises of infinite systems, random distributions, and the absence of spatial correlation [20–22]. With microtomography data, assumptions on the pore network can be verified or rejected. Site and bond models have both been employed for porous media applications. Bond models are considered to be more appropriate since bonds may be analogous to pore throats which control fluid motion in capillary-dominated systems [23]. Identification of pore throats and pore bodies in images such as

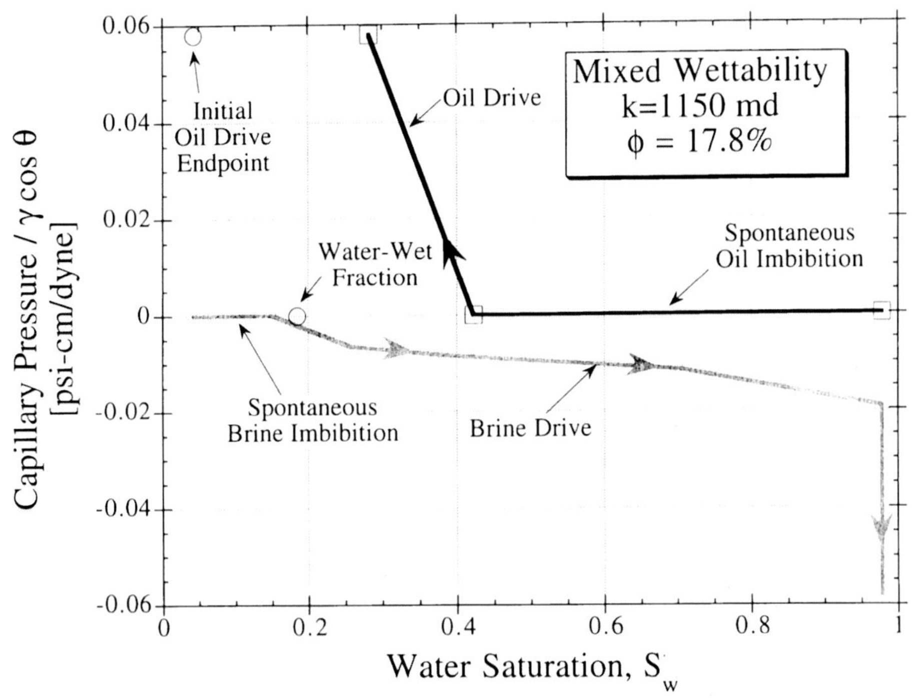

Fig. 8. Combined USBM–Amott test simulation on mixed-wet Berea sandstone with 18% of the pore system designated as water-wet. USBM index, $W = -0.33$; Amott indices: $\delta_w = 0.12, \delta_o = 0.80, I = -0.68$.

those in Figure 5 is not at all straightforward. In this work, the connectivity of blobs of like phase occupancy is examined directly without regard for a model framework.

Predicted fluid spatial distributions were examined for isolated fluid segments. Independent clusters of voxels were examined for their size distribution and ability to support flow. The range of voxels covered by the largest cluster was of particular interest for percolation testing on finite size networks. Figure 9 shows the lateral extent of the largest cluster for both the wetting and nonwetting phases as a function of the inverse pore entry pressure, which has been recast in terms of the Laplace equation variables. The average cluster extent from the three principal directions was used since this Berea sample was relatively isotropic. What is observed is striking in that at no point do both fluid phases percolate simultaneously. The interpretation, however, should be made in light of unresolved microporosity and film flow contributions for the wetting phase which are assumed to always maintain connectivity for that phase.

The implications concerning the absence of a large bicontinuous region are still far reaching in terms of flow behavior for the nonwetting phase. Over the majority of the saturation range, slug flow is predicted in that wetting and nonwetting

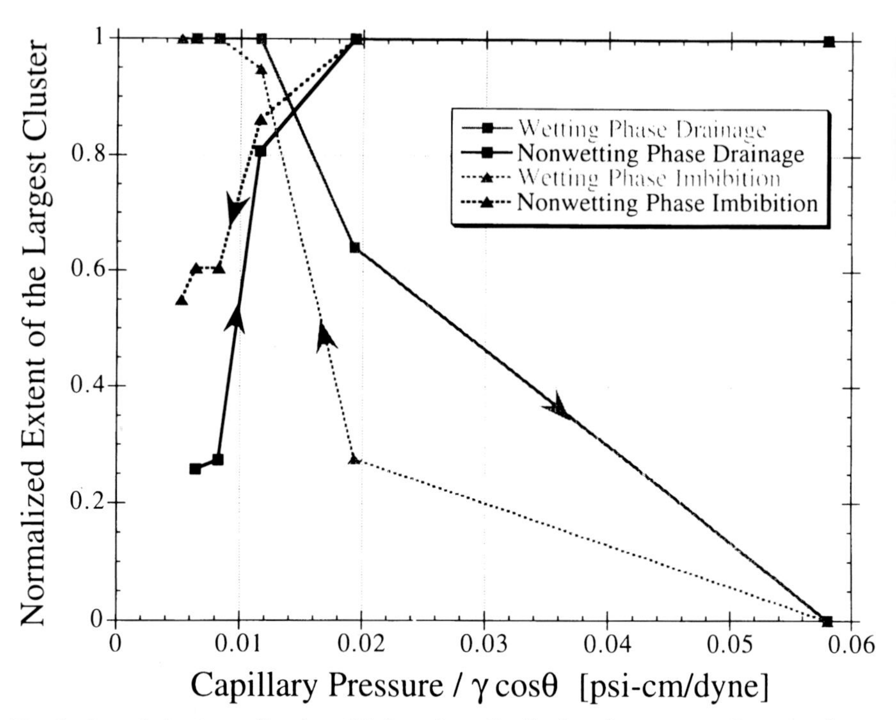

Fig. 9. Percolation in predicted equilibrium phase distributions for water-wet Berea sandstone in the absence of thin-film connectivity. The average extent over the three principal directions of the largest cluster for each phase is shown as a function of Laplace equation variables.

phases can not flow in completely independent paths in an equilibrium sense. Slug flow would not be appropriately represented by a simple extension of Darcy's Law for single-phase systems in that pressure drop and flow rate need not be proportional. Similar observations concerning percolation properties have been made with additional producing reservoir microtomography samples.

Additional factors should be considered. Higher resolution imaging may allow some two-phase mutual connectivity to be observed. More heterogeneous, non-isotropic specimens may actually help maintain mutual phase connectivity. Also, film contributions could be extremely important as they could delay the onset of slug flow for some saturation ranges. Mixed wettability states might also be significant in that each phase can then have a film flow contribution to maintain connectivity. Another interpretation might preclude observation of certain saturation ranges except in transient flow conditions. At issue in part is the difference between steady and unsteady state relative permeability measurements.

6. Transport Properties

The equilibrium fluid distributions predicted from this work can provide the basis for transport property prediction. As an example, an oil–water distribution for Berea

TABLE I. Experimental data on oil permeability in a Berea sandstone as a function of immobile water saturation

Capillary pressure (psi)	Immobile water saturation (%)	Permeability to oil (μm^2)
10	45.6	0.422
35	26.8	0.634
80	20.6	0.694

was furnished to a Lattice Boltzmann technique [24] for fluid flow modeling. This choice of fluid distribution was representative of a low initial water saturation of 18%. Simulation was performed on a $256 \times 256 \times 256$ representation of the Berea pore system. The oil phase was injected at low capillary number conditions until pressure and velocity profiles appeared stable. The permeability of each phase was computed, yielding near end-point permeabilities of 0.45 and 0.01 μm^2 for the oil and water phases respectively. Although no direct laboratory measurements were performed for a sample of this size, the end-point oil permeability is in reasonable agreement with a large number of core-scale measurements performed on Berea sandstones with similar porosity and single phase permeability [19].

We have since carried out mineral oil permeability measurements at low stationary brine saturations on a 2 inch diameter by 4 inch long piece of the parent material from which the microtomography sample was taken. The results are provided in Table I. A porous plate technique was used to establish a uniform, equilibrium saturation prior to oil flow measurement. Laboratory tests indicated 23% porosity rather than the 18% in our thresholded images and digital representations for simulation. The difference is attributed to roughness and microporosity below the 10-μm cutoff. Thus, all simulated saturations were computed against a constant background of 22% water saturation. The corrected endpoint simulation saturation was, therefore, 40%. Interpolating between tabular values, the laboratory oil permeability for comparison with simulation is found to be 0.49 μm^2, in unexpectedly close agreement with simulation on a sample 10^5 times smaller.

7. Limitations

It should be noted that the prediction of equilibrium fluid distributions utilizes an assemblage of inscribed spheres. For the sphere, the two principal radii of curvature are equal and positive. The mean curvature of convex shapes, which are important in imbibition, is not extracted. Thus, there may be a tendency to over-estimate phase trapping in the network model simulation at the higher capillary pressures during imbibition. A more rigorous solution is not perceived at this time. Also, the spheres are tangent to the solid, indicating perfect wetting. Although results are scalable for comparison with actual data in terms of interfacial tensions, nonzero contact angles are not. Since saturations obtained would differ principally by the volume associated with menisci, the appropriately scaled results should be approximate for arbitrary contact angle. Extension of the algorithm to consider sphere segments which intersect walls is nontrivial and is beyond the scope of this investigation.

This development considers fluid movement as the result of capillary pressure changes in the limit of zero capillary number. High velocity or viscous-dominated flows are expected to give different fluid distributions. This is a topic of future work.

The topic of representative element volumes was not addressed. The network modelling described can be applied to much larger systems. The representativeness issue then becomes a question of data acquisition.

8. Conclusions

A network model capable of simulating capillary-dominated displacements which result from increases or decreases in capillary pressure has been developed. Equilibrium fluid distributions can be derived on complex network geometries, such as those obtained from micro-imaging techniques on physical reservoir rock specimens. The network model algorithm can handle mixed wettability scenarios and gives reasonable results for wettability indices. The initial observations concerning percolation properties on imaged porous rocks have strong implications for modeling multiphase flow in that slug flow regimes are predicted in steady state two-phase flow. Transport properties can be estimated using predicted equilibrium fluid distributions. Higher resolution and larger volume samples from micro-imaging are desirable.

Acknowledgements

The microtomography data on Berea sandstone was graciously made available by Jay Jasti. Discussions with H. M. Princen, R. N. Vaidya, M. M. Honarpour, and D. D. Huang on the porosimetry simulation are gratefully acknowledged. The Lattice Boltzmann simulation technique used herein was developed at Los Alamos National Laboratory by Shiyi Chen, Daryl Grunau, and Ken Eggert. The Lattice Boltzmann simulation was performed with the CM5 computational resources of the Advanced Computer Laboratory at Los Alamos National Laboratory.

References

1. Fatt, I.: 1956, The network model of porous media, I. Capillary pressure characteristics, *Petroleum Trans. AIME* **207**, 144–159.
2. Mohanty, K. and Salter, S.: 1982, Multiphase flow in porous media, I. Pore-level modeling, SPE paper 11018, Proc. 57th Annual Fall Technical Conf. and Exhibition Soc. Petroleum Engineers of AIME, New Orleans, September 26–29.

3. Lin, C-Y. and Slattery, J. C.: 1982, Three-dimensional, randomized, network model for two-phase flow through porous media, *AIChE J.* **28** (2), 311–324.
4. Dias, M. M. and Payatakes, A. C.: 1986, Network models for two-phase flow in porous media, Part 1. Immiscible microdisplacement of non-wetting fluids, *J. Fluid Mech.* **164**, 305–336.
5. McDougall, S. R. and Sorbie, K. S.: 1992, Network simulations of flow processes in strongly wetted and mixed-wet porous media, *Proc. 3rd European Conf. Mathematics of Oil Recovery*, Delft, June 17–19, pp. 169–181.
6. Dunsmuir, J. H., Ferguson, S. R., D'Amico, K. L. and Stokes, J. P.: 1991, X-ray microtomography: A new tool for the characterization of porous media, SPE paper 22860, Proc. 66th Annual Technical Conf. and Exhibition Soc. Petroleum Engineers, Dallas, October 6–9.
7. Jasti, J., Jesion, G., and Feldkamp, L.: 1990, Microscopic imaging of porous media using X-ray computer tomography, SPE Paper 20495, Proc. 65th Annual Technical Conf. and Exhibition of the Soc. Petroleum Engineers, New Orleans, September 23–26.
8. Coles, M. E., Spanne, P., Muegge, E. L. and Jones, K.: 1994, Computed microtomography of reservoir core samples, Proc. Int. Symp. Soc. Core Analysts, Stavanger, September 12–14.
9. Adler, P. M., Jacquin, C. G. and Quibler, J. A.: 1990, Flow in simulated porous media, *Int. J. Multiphase Flow* **16** (4), 691–712.
10. Adler, P. M., 1992, *Porous Media: Geometry and Transports*, Butterworth/Heinemann, Stoneham, MA.
11. Spanne, P., Thovert, J. F., Jacquin, C. J., Lindquist, W. B., Jones, K. W. and Adler, P. M.: 1994, Synchrotron computed microtomography of porous media: Topology and transports, *Phys. Rev. Lett.* **73** (14), 2001–2004.
12. Anderson, W. G.: 1986, Wettability literature survey, Part 2: Wettability measurement, *J. Petroleum Technol.* November, 1246–1262.
13. Hazlett, R. D.: 1990, Fractal applications: Wettability and contact angle, *J. Colloid Interface Sci.* **137** (2), 527–533.
14. Hazlett, R. D.: 1992, On surface roughness effects in wetting phenomena, *J. Adhesion Sci. Technol.* **6** (6), 625–633.
15. Fassi-Fihri, O. and Robin, M. and Rosenberg, E.: 1991, Wettability studies at the pore level: A new approach by the use of cryo-scanning electron microscopy, SPE paper 22596, Proc. 66th Annual Technical Conf. and Exhibition Soc. Petroleum Engineers, Dallas, October 6–9, 97–110.
16. Melrose, J. C.: 1982, Interpretation of mixed wettability states in reservoir rocks, SPE paper 10971, Proc. 57th Annual Fall Technical Conf. and Exhibition Soc. Petroleum Engineers of AIME, New Orleans, September 26–29.
17. Blunt, M., King, M. and Scher, H.: 1992, Simulation and theory of two-phase flow in porous media, *Phys. Rev. A* **46** (12), 7680–7699.
18. Adamson, A. W.: 1967, *Physical Chemistry of Surfaces*, Interscience, New York, pp. 548–549.
19. Jadhunandan, P. P.: 1990, Effects of brine composition, crude oil, and aging conditions on wettability and oil recovery, Ph. D. Thesis, New Mexico Institute of Mining and Technology.
20. Shante, V. K. S. and Kirkpatrick, S.: 1971, An introduction to percolation theory, *Adv. in Phys.* **20**, 325–357.
21. Essam, J. W.: 1980, Percolation theory, *Rep. Prog. Phys.* **43**, 833–912.
22. Stauffer, D., Scaling theory of percolation clusters, *Phys. Rep.* **54** (1), 1–74.
23. Chandler, R., Koplik, J., Lerman, K. and Willemsen: 1982, Capillary displacement and percolation in porous media, *J. Fluid Mech.* **119**, 249–267.
24. Grunau, D., Chen, S. and Eggert, K.: 1993, A lattice Boltzmann model for multiphase fluid flows, *Phys. Fluids A* **5** (10), 2557–2562.

Transport in Porous Media **20:** 37–76, 1995.

Surface Tension Models with Different Viscosities

I. GINZBOURG* and P. M. ADLER
LPTM, Asterama 2, Avenue du Téléport, 86360 – Chasseneuil, France

(Received: May 1994)

Abstract. Surface tension in ILB models for fluids with different viscosities and different numbers of rest populations is derived, starting from the so-called mechanical definition. It is shown that the standard perturbation, inserted into these models in order to create surface tension, should be slightly modified for models with different viscosities in order to avoid the dependence of surface tension upon the actual phase distribution. The analytical results are numerically confirmed by mechanical and bubble tests. It is demonstrated also that the perturbation of the lattice Boltzmann equation gives rise to the appearance of anisotropic terms in population solutions related to anomalous currents and density fluctuations. When particular values of the eigenvalues of the collision operators are used, these spurious currents are annihilated in the time-independent solutions of the mechanical tests in arbitrarily inclined channels when bounce-back conditions are imposed at the solid boundaries.

Key words: surface tension, viscosity, perturbation, lattice gas model, lattice-Boltzmann model, immiscible lattice gas model.

1. Introduction

During the last few years, numerous models based on single phase *Lattice Gas* (LG) and *Lattice Boltzmann* (LB) models have been developed. They originated in the FHP 2D *Lattice Gas* automation introduced by Frisch *et al.* (1986, 1987) and in the FCHC 4D *Lattice Gas* developed by d'Humières *et al.* (1986) for simulations in three dimensions. The assumption of molecular chaos is made in *Lattice Boltzmann* models suggested by McNamara and Zanetti (1988) and developed by Higuera and Jimenez (1989). Incompressible Navier–Stokes equations are recovered by both approaches in an asymptotic limit.

Some of these single-phase methods have been extended to simulate the behaviour of two or more immiscible fluids. These algorithms may be listed as follows: the *Immiscible Lattice Gas* (ILG) multiphase models introduced by Rothman and Keller (1988), Gunstensen and Rothman (1991); the various LG approaches to study Saffman–Taylor type instabilities summarized by Hayot (1991); the alternative to ILG lattice gas models of Rem and Somers (1989); the extension of the ILG to Boltzmann models due to Gunstensen *et al.* (1991) and Grunau (1993); finally, the recent *liquid–gas* models with phase transition developed for both Lattice Gas (Appert and Zaleski, 1990, 1993; Appert *et al.*, 1991) and Lattice Boltzmann (Appert, 1993) approaches.

---

* Present address: ASCI, CNRS, Bât. 506, Université Paris Sud, 91405 Orsay Cedex, France.

The main idea of ILG models consists of the introduction of two species of particles in such a manner that in regions occupied by species of one type (color), ILG is identical with standard single LG method; however, in nodes occupied by particles of both colours, the collision results depend on the phase distribution in the neighbouring nodes. Such a modified collision operator provides the surface tension.

The models of Rem and Somers (1989) are rather similar, but now the holes also carry information about the fluid colour in a site. Consequently, the interfacial collision operator uses only the local-site information about phase distribution, and thus it obeys the basic principles of lattice gas methods.

Immiscible Lattice Boltzmann (ILB) multiphase models are developed on the basis of the single-phase LB models and ILG models; they consist of Boltzmann equations supplemented by perturbation of populations near the interface in order to introduce surface tension; the separation of phases is performed in the same manner as in ILG models. The multiphase methods of Grunau (1993) are essentially the same, but they follow the *Pressure-Corrected* LBE (PCLBE) single-phase model introduced by Chen (1992).

Thus, in the ILG and ILB models, there are at least two kinds of particles which can be separated; this separation yields two (or more) immiscible phases of the same density; in contrast with it, in the *liquid–gas* models there is only one type of particles, but they interact in some manner; a phase-separation transition into light and dense phases is obtained. These interactions are similar to attractive or repulsive forces; they are implemented in *liquid–gas* models by the exchange of some quantity of momentum between sites located at a certain distance one from another; the total momentum conservation law is not violated by these interactions.

All these models show a qualitatively adequate interfacial behaviour. Nevertheless, the precision of the methods, their abilities to satisfy the usual macroscopic interfacial conditions (for example, Edwards *et al.*, 1991) for tangential directions are not thoroughly discussed. Moreover, the appearance of anomalous currents near an inclined interface has been reported twice by Gunstensen (1992) and by Appert (1993); no successful remedy could be found so far. The analysis of the interfacial conditions imposed by two generalised ILB models 1 and 2 is given by Ginzbourg and Adler (1994b) for plane interfaces; the major results of this analysis are gathered in Table I. The main steps of these two immiscible lattice Boltzmann models used in our theoretical and numerical study are briefly recalled in Section 2.

The interfacial behaviour of the immiscible lattice models described by the Laplace law in the direction normal to the interface has been much more studied. The attempts to predict the surface tension σ could be divided into two general approaches. The first one due to Gunstensen *et al.* (1991) is used by Chen (1991) and Adler *et al.* (1994) for ILG, by Appert (1993) for *liquid–gas* and by Grunau (1993) for two-phase PCLBE; in these works, surface tension is derived from the *mechanical definition* given by Rowlinson and Widom (1982) which connects σ

with the difference between the normal and tangential components of the pressure tensor which becomes anisotropic near the interface. These components are derived from the solutions of the model equations; although the basic equations of the models are not the same (see the Boltzmann equations with perturbation in ILB models or the Boltzmann equations with an interaction operator in *liquid–gas*), essentially the same constitutive condition for population solutions is used to extract σ from the mechanical test. This assumption actually neglects density fluctuations and anomalous currents which appear in these models due to the modifications of lattice-Boltzmann equations. The resulting expressions for σ in the ILB models depend on the equilibrium bulk density and on the kinematic viscosity (averaged for different bulk viscosities by Grunau, 1993); for the *liquid–gas* models, it is not possible at the moment to prescribe the densities at equilibrium; hence numerical evaluations of the equilibrium densities are necessary to calculate σ. This assumption about population solutions is not used by Adler *et al.* (1994); a time-independent solution for populations in the interfacial layer is derived from the macroscopic equations using the Boltzmann approximation and boundary conditions are introduced at the ends of the interfacial layer.

This mechanical definition of σ is then numerically verified by the Laplace law for a bubble; this is the so-called bubble test.

In the second approach, used by Burgess *et al.* (1988) and Somers and Rem (1991), surface tension is tentatively derived from the Laplace law itself. Equilibrium solutions for populations are introduced by Somers and Rem (1991) using the FHP lattice properties and a symmetry of the 'bubble test'; then the components of the pressure tensor are derived from these solutions leading to the relation between the pressure difference and the bubble radius. Thus, σ could be expressed in terms of some coefficients of equilibrium solutions; these coefficients were supported by numerical experiments.

In Section 3 of this paper, we define the stationary (time-independent) populations solutions in mechanical tests for both ILB models with different viscosities and rest populations. No assumption about the populations is needed; the Boltzmann equation with perturbation is analytically solved for an arbitrarily inclined 2D configuration of limited extent (Figures 2a and c) using bounce-back conditions on the solid boundaries. Due to the special choice of the eigenvalues, the spurious currents can be annihilated in time-independent solutions for an arbitrarily inclined 2D configuration of limited extent; however, it turns out that it is only in the particular case of an interface inclination at an angle $\theta = 45°$, that these currents can be eliminated in a periodical configuration (Figures 2b and d). When the spurious currents are suppressed, the population solutions allow the derivation of σ from the mechanical definition in terms of the perturbation function and of the eigenvalue of the collision operators which determines the kinematic viscosity. Then, the values of σ are calculated for two principal types of interfaces using the obtained general formula. First, an interface is implicity tracked between two rows of lattice nodes with different predominant phases; this corresponds to Model

1 which uses the rule of majority of colour at the interface (Figure 1a). Second, an interface is explicity located in some lattice nodes which separate one phase from another (Figure 1b); this corresponds to Model 2 with a special interfacial collision matrix. Taking into account the density fluctuations, the results of Gunstensen and Rothman (1992), Gunstensen (1992) found that for the particular case of the fluids with equal viscosities and equal numbers of rest populations, they are more accurately redefined. The resulting populations also allow the validity of the constitutive condition mentioned above to be determined.

Section 4 is devoted to numerical simulations which verify the theoretical predictions for the surface-tension coefficient. Anomalous currents are shown to disappear and time-independent solutions are obtained by an adequate choice of the eigenvalues. The influence of the anomalous density fluctuations on the macroscopic solutions is also demonstrated.

Some possible extensions of this paper are tentatively discussed in Section 5.

2. Basic ILB Models 1 and 2

Let us briefly present the two immiscible two-phase 3D models with rest populations which are going to be used. The interface is defined as the set of nodes where the two phases are simultaneously present. The first model does not provide any new procedure, since it was already developed by Gunstensen and Rothman (1992), but the possibility of having different viscosities as well as different numbers of rest populations is formally incorporated into it; at the interface where both populations are present, the collision matrix is chosen by the majority rule. Hence, the choice of the collision matrix is ambiguous at the nodes where the red mass is equal to the blue one. The second model is the same in bulk, but a special collision matrix is introduced at the interface in order to satisfy the standard interfacial conditions at first order. Both models are schematised in Figures 1a and b; the general results of the analysis are gathered in Table I.

Let the populations of red or blue phase of velocity $\mathbf{C}_i$ be $N_i^R(\mathbf{r},t)$ or $N_i^B(\mathbf{r},t)$, ($i = 0,\dots,b_m = 24$), respectively. The index 0 corresponds to rest populations; the numbers of rest populations are L^R or L^B for red or blue phase, respectively; the moving populations may have $b_m = 24$ velocities $\mathbf{C}_i$ in the FCHC model. The algorithm consists of the following five steps which are now described.

2.1. DEFINITION OF THE FIELDS AT TIME t

2.1.1. *Calculation of the Densities of Each Phase ρ_R, ρ_B*

$$\rho_K(\mathbf{r},t) = \sum_{i=0}^{24} N_i^K(\mathbf{r},t) W_i^K, W_0^K = L^K,$$

$$W_i^K = 1, i = 1,\dots,24 \quad \text{for} \quad K = R,\ B. \tag{2.1.a}$$

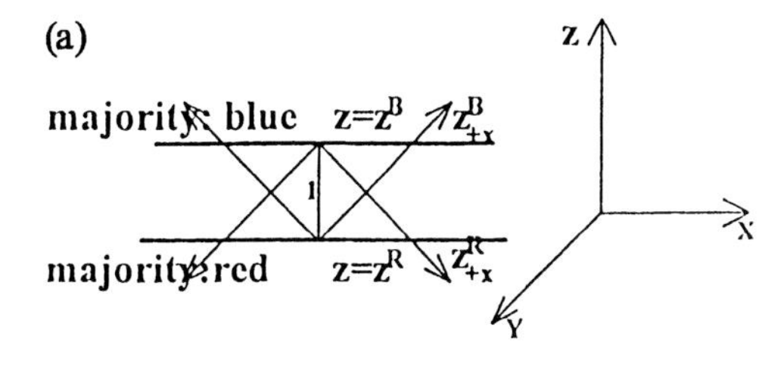

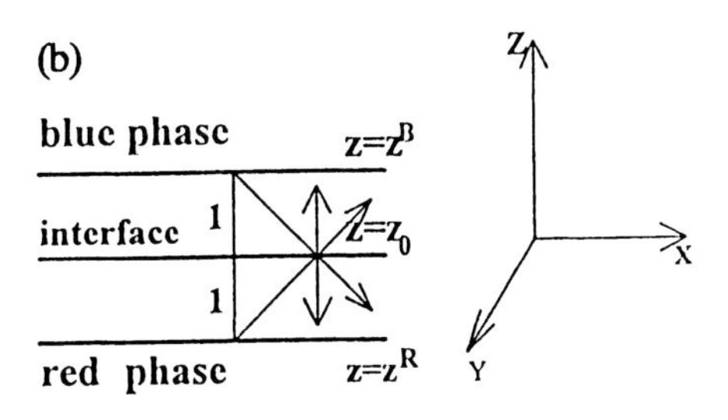

Fig. 1. Plane interface between blue and red fluids in Model 1(a) and in Model 2(b).

2.1.2. *Calculation of the Total Population N_i and of the Total Density ρ*

$$N_i(\mathbf{r},t)W_i = N_i^R W_i^R + N_i^B(\mathbf{r},t)W_i^B, \quad i = 0,\dots,24,$$

$$W_0 = L(\text{cf. (2.2.b) and (2.2.c)}), \quad W_i^K = 1, \quad i = 1,\dots,24,$$

$$\rho(\mathbf{r},t) = \rho_R(\mathbf{r},t) + \rho_B(\mathbf{r},t). \tag{2.1.b}$$

2.1.3. *Addition of External Forces $\mathbf{F}^R$ and $\mathbf{F}^B$ according to the single-phase Algorithm of Succi et al. (1989)*

$$N_i'(\mathbf{r},t) = N_i(\mathbf{r},t) + \delta N_i^F(\mathbf{r},t);$$

$$\delta N_i^F(\mathbf{r},t) = \frac{D}{c^2 b_m}\rho(\mathbf{r},t)(\mathbf{F}(\mathbf{r},t)\cdot\mathbf{C}_i,$$

$$D = 4, \ \forall i\, c^2 = (\mathbf{C}_i\cdot\mathbf{C}_i) \overset{\text{FCHC}}{\equiv} 2, \tag{2.1.c}$$

where the expression of $\mathbf{F}(\mathbf{r},t)$ depends upon the chosen model:
Model 1: by majority of colour

$$\mathbf{F}(\mathbf{r},t) = \begin{cases} \mathbf{F}^R, \rho_B(\mathbf{r},t) < \rho_R(\mathbf{r},t), \\ \mathbf{F}^B, \rho_B(\mathbf{r},t) > \rho_R(\mathbf{r},t). \end{cases} \tag{2.1.d}$$

Model 2:

$$\mathbf{F}(\mathbf{r},t) = \frac{\mathbf{F}^R\rho_R(\mathbf{r},t) + \mathbf{F}^B\rho_B(\mathbf{r},t)}{\rho(\mathbf{r},t)}. \tag{2.1.e}$$

TABLE I. Results for recolouring algorithm: plane stationary interface

		Model 1 (Fig. 1.a)		Model 2 (Fig. 1.b)	
		Choice of the interfacial matrix by majority of colour: $A_{ij} = \begin{cases} A^R_{ij}(\lambda^R_\psi, \lambda^R_2, L^R), \rho_R > \rho_B \\ A^B_{ij}(\lambda^B_\psi, \lambda^B_2, L^B), \rho_B > \rho_R \end{cases}$ No conditions on $\lambda^{\text{int.}}_\psi$: $\lambda^{\text{int.}}_\psi(z^B) = \lambda^B_\psi$ $\lambda^{\text{int.}}_\psi(z^R) = \lambda^R_\psi$		Introduction of an additional collision matrix $\mathbf{A}^{\text{int}}$ at interface $z = z_0$ $A^{\text{int.}}_{ij}(\lambda^{\text{int.}}_\psi, \lambda^{\text{int.}}_2, L^{\text{int.}})$ with $\lambda^{\text{int.}}_\psi = 2\frac{\lambda^R_\psi + \lambda^B_\psi + \lambda^R_\psi \lambda^B_\psi}{(\lambda^R_\psi + \lambda^B_\psi + 4)}$ then $\nu^{\text{int.}} = 2\frac{\nu^R \nu^B}{\nu^R + \nu^B}$	
Order	Flow conditions	Resulting interfacial conditions for tangential velocity u_α , $\alpha = \{x, y\}$			
		$L^R = L^B$	$L^R \neq L^B$	$L^R = L^B$	$L^R \neq L^B$
ε	Different viscosities: $\nu^R \neq \nu^B$	$u^R_\alpha = u^B_\alpha$	$\rho_R u^R_\alpha = \rho_B u^B_\alpha$	$u^R_\alpha = u^B_\alpha$	$\rho_R u^R_\alpha = \rho_B u^B_\alpha$
	Identical body forces: $\rho_R F^R = \rho_B F^B$	$\rho_R \nu^R \frac{\partial u^R_\alpha}{\partial z} = \rho_B \nu^B \frac{\partial u^B_\alpha}{\partial z}$			
	Interfacial conditions[a]	$z^{\text{int.}} = \frac{z^R + z^B}{2}$		$z^{\text{int.}} = z_0$	
	Different body forces: $\rho_R F^R \neq \rho_B F^B$	$z^{\text{int.}} \neq \frac{z^R + z^B}{2}$		$z^{\text{int.}} \neq z_0$	
ε^2	Two-phase Poiseuille flow with $F^R \neq F^B$ $\nu^R \neq \nu^B$ $L^R = L^B$	Exact solution if $\lambda^R_2 = \lambda_{2c}(\lambda^R_\psi)$ $\lambda^B_2 = \lambda_{2c}(\lambda^B_\psi)$ $\lambda_{2c} = -8\frac{(\lambda_\psi + 2)}{(\lambda_\psi + 8)}$		Exact solution if $\lambda^R_2 = \lambda^B_2 = \lambda^{\text{int.}}_2 = \lambda^{(0)}_2$ $\lambda^{(0)}_2 = -4\frac{(\lambda_\psi + 2)}{(4 - \lambda_\psi)}$ and $\rho^R_I(z_0) = \rho^B_I(z_0)$	

[a] $C^{2R}_s \rho_R = C^{2B}_s \rho_B$; $u_z = 0$; $\frac{\partial u_\alpha}{\partial \alpha} = 0, \alpha \in \{x, y\}$.

2.1.4. *Calculation of the Total Velocity* u

$$\mathbf{u}(\mathbf{r}, t) = \frac{\sum_{i=1}^{24} N'_i(\mathbf{r}, t)\mathbf{C}_i}{\rho(\mathbf{r}, t)} - \tfrac{1}{2}\mathbf{F}(\mathbf{r}, t). \tag{2.1.f}$$

Addition of $\{-\frac{1}{2}\mathbf{F}\}$ to the classical velocity expression is caused by the presence of the term $\delta N_i^F(\mathbf{r},t)$ given by (2.1.c) in the lattice Boltzmann equation (see Ginzbourg and Adler, 1994a,b). Defined in such a manner, the velocity $\mathbf{u}(\mathbf{r},t)$ corresponds to the local velocity in the equilibrium solutions for populations, and consequently, to the macroscopic velocity of the simulated fluid.

2.2. COLLISION OF TOTAL POPULATIONS

$$N_i^*(\mathbf{r},t) = N_i'(\mathbf{r},t) + \sum_{j=0}^{24} A_{ij} N_j^{\text{neq.}}(\mathbf{r},t) W_j, \quad i = 0,\ldots,24, \tag{2.2.a}$$

$$N_i^{\text{neq.}} = N_i' - N_i^{\text{eq.}}; \quad W_0 = L, \quad W_i = 1, i = 1,\ldots,24,$$

where the equilibrium solution $\mathbf{N}^{\text{eq.}}$ is known (Ginzbourg and Adler, 1994a). The collision operator $\mathbf{A}(\mathbf{r},t)$ and the number of rest populations $L(\mathbf{r},t)$ are different in Models 1 and 2.

Model 1

$$L = \begin{cases} L^R, & \rho_R(\mathbf{r},t) > \rho_B(\mathbf{r},t), \\ L^B, & \rho_R(\mathbf{r},t) < \rho_B(\mathbf{r},t); \end{cases}$$

Model 2

$$L = \begin{cases} L^R, & \rho_B(\mathbf{r},t) = 0 \\ L^{\text{int}}, & \rho_R(\mathbf{r},t) \neq 0, \text{ and } \rho_B(\mathbf{r},t) \neq 0 \\ L^B, & \rho_R(\mathbf{r},t) = 0; \end{cases}$$

$$A_{ij} = \begin{cases} A_{ij}^R, & \rho_R(\mathbf{r},t) > \rho_B(\mathbf{r},t), \\ A_{ij}^B, & \rho_R(\mathbf{r},t) < \rho_B(\mathbf{r},t); \end{cases} \tag{2.2.b}$$

$$A_{ij} = \begin{cases} A_{ij}^R, & \rho_B(\mathbf{r},t) = 0, \\ A_{ij}^{\text{int.}}, & \rho_R(\mathbf{r},t) \neq 0, \text{ and } \rho_B(\mathbf{r},t) \neq 0, \\ A_{ij}^B, & \rho_R(\mathbf{r},t) = 0; \end{cases} \tag{2.2.c}$$

The condition of mass conservation imposed on the matrix A can be expressed by (see A.1)

$$\sum_{j=0}^{24} A_{ij} W_j = 0, \quad W_0 = L, \quad W_i = 1, \quad i = 1,\ldots,24 \tag{2.2.d}$$

Consequently, when the matrix $A(\mathbf{r},t-1)$ has to be replaced by some different matrix $A(\mathbf{r},t)$, the value of rest population $N_0'(\mathbf{r},t)$ has to be recalculated before collision in order to conserve mass during the collision, if the numbers of rest populations associated with these matrices are different

$$N'_0(\mathbf{r},t) = \frac{N_0(\mathbf{r},t)L(\mathbf{r},t-1)}{L(\mathbf{r},t)}. \tag{2.2.e}$$

Each bulk kinematic viscosity ν^k is defined by the eigenvalue λ_ψ^K of the corresponding collision matrix

$$\nu^K = -\frac{c^2}{D+2}\left(\frac{1}{2}+\frac{1}{\lambda_\psi^K}\right) \quad \text{for } K = R,\ B. \tag{2.2.f}$$

2.3. SURFACE TENSION

A perturbation $\delta N_i(\mathbf{r},t)$ of the populations creates surface tension

$$N''_i(\mathbf{r},t) = N^*_i(\mathbf{r},t) + \delta N_i(\mathbf{r},t),$$

$$\delta N_i(\mathbf{r},t) = C^{\text{per.}}\left\{\frac{(\mathbf{C}_i\cdot\mathbf{f})^2}{(\mathbf{f}\cdot\mathbf{f})^2} - \frac{1}{2}\right\}, \quad i = 1,\ldots,24,\ \delta N_0(\mathbf{r},t) \equiv 0, \tag{2.3.a}$$

where $C^{\text{per.}}$ is some function of $\mathbf{f}(\mathbf{r},t)$; $C^{\text{per.}}(\mathbf{r},t)$ decreases when the distance from the interface increases. A standard form for $C^{\text{per.}}(\mathbf{r},t)$ has been given by Gunstensen (1992)

$$C^{\text{per.}}(\mathbf{r},t) = A^{\text{per.}}|\mathbf{f}(\mathbf{r},t)|, \tag{2.3.b}$$

where $A^{\text{per.}}$ is an arbitrary constant. The local colour gradient $\mathbf{f}(\mathbf{r},t)$, introduced by Rothman and Keller (1988), is defined by

$$\mathbf{f}(\mathbf{r},t) = \sum_{i=1}^{24}\mathbf{C}_i[\rho_R(\mathbf{r}+\mathbf{C}_i,t) - \rho_B(\mathbf{r}+\mathbf{C}_i,t)]. \tag{2.3.c}$$

2.4. RECOLOURING

The total population is recoloured in order to separate the populations $N_i''^R$ and $N_i''^B$ before subsequent propagation; $N_i''^R(\mathbf{r},t)$ is set equal to a value which maximises the function $\vartheta(N_i^R)$

$$\vartheta(N_i^R) = \sum_{i=1}^{24} N_i^R\mathbf{C}_i\cdot\mathbf{f}(\mathbf{r},t), \tag{2.4.a}$$

together with the constraints

$$\sum_{i=0}^{24} N_i^R W_i^R = \rho_R(\mathbf{r}, t), \tag{2.4.b}$$

$$0 \leqslant N_i^R W_i^R \leqslant N_i''(\mathbf{r}, t) W_i(\mathbf{r}, t). \tag{2.4.c}$$

In order to conserve the total momentum in a site, $N_i''^B(\mathbf{r}, t)$ is given by

$$N_i''^B W_i^B = N_i'' W_i - N_i''^R W_i^R, \quad i = 0, \ldots, 24. \tag{2.4.d}$$

2.5. PROPAGATION

$$N_i^K(\mathbf{r} + \mathbf{C}_i, t+1) W_i^K = N_i''^K(\mathbf{r}, t) W_i^K,$$
$$i = 0, \ldots, 24 \qquad \text{for } K = R,\ B.$$

Then the iteration procedure 2.1 – 2.5 is repeated.

Alternative ILB models with different viscosities are discussed by Grunau (1993) for two-phase PCLBE models with the diagonalized collision operator, which is completely defined by a single constant, the so-called *relaxation parameter* τ; the interfacial solution for τ is constructed in order to continuously connect the bulk relaxation parameters; no theoretical study on the precision of models with such interfacial collision operators has been reported yet to the best of our knowledge.

3. Analysis of Surface-Tension Coefficients

3.1. GENERAL

For the ILG models on the FHP lattice (Adler *et al.*, 1994), the FHP ILB (Gunstensen *et al.*, 1991) and the FCHC ILB models (Gunstensen, 1992), the PCLBE 3D model (Grunau, 1993) and for the FHP *liquid–gas* Boltzmann model (Appert, 1993) the same scheme of derivation of the surface tension coefficient is proposed. Namely, surface tension may be defined by the relation, which is called the *mechanical definition* (Rowlinson and Widom, 1982)

$$\sigma = \int_{-\infty}^{+\infty} (p_n - p_t)\, \mathrm{d}n, \tag{3.1}$$

where p_n and p_t are the normal and tangential components of the pressure tensor and n is the coordinate perpendicular to the interface. This means that near the interface the pressure tensor is anisotropic

$$\mathbf{P} = \begin{pmatrix} p_t & & \\ & p_t & \\ & & p_n \end{pmatrix}. \tag{3.2.a}$$

For ILG and ILB models, assuming that both fluids are at rest, the components p_n and p_t are given by Gunstensen *et al.* (1991)

$$p_n = \sum_{i=1}^{b_m} N_i C_{in}^2, \qquad p_t = \sum_{i=1}^{b_m} N_i C_{it}^2, \tag{3.2.b}$$

where C_{in} and C_{it} are the normal and tangential components of the velocity. These relations are slightly different for *liquid–gas* due to the presence of the interaction operator in the lattice Boltzmann equation which modifies the local momentum (Appert, 1993).

Let us assume that a colour gradient $\mathbf{f}(\mathbf{r}, t)$ is perpendicular to the interface with negligible deviations due to the discretization from the normal direction. The normal and tangential components of the velocity can be expressed by means of the unit vector $\mathbf{f}$, in the direction of the colour gradient in the form

$$\mathbf{C}_{in} = \mathbf{C} \cdot \mathbf{f}_e \mathbf{f}_e, \qquad \mathbf{C}_{it} = \mathbf{C}_i \cdot (\mathbf{I} - \mathbf{f}_e \mathbf{f}_e) \tag{3.2.c}$$

where

$$\mathbf{f}_e \overset{\text{def.}}{=} \frac{\mathbf{f}}{|\mathbf{f}|} \overset{\text{def.}}{=} (f_{e_x}, f_{e_y}, f_{e_z}, 0) \tag{3.2.d}$$

and 'def.' means *definition*.

Thus, the mechanical definition (3.1) takes a form

$$\sigma = \int_{-\infty}^{+\infty} \left\{ \sum_{i=1}^{b_m} N_i(n) \Phi_i(n) \right\} \mathrm{d}n, \quad \Phi_i \overset{\text{def.}}{=} (C_{in}^2 - C_{it}^2). \tag{3.3}$$

Hence, one has to determine the populations solutions N_i in the configurations which obey the conditions of the mechanical definition (3.1); the analytical or numerical generation of such solutions is called the *mechanical test*. A system of two fluids contained in a box, limited by solid walls which are parallel with an interface and with boundary conditions periodic in the x and y directions, is usually considered (Gunstensen, 1992; Appert, 1993). The fluids are separated by an interface whose normal makes an angle θ with the z-axis. Examples of such a cell, called E_s, are given in Figures 2.a and c.

The other box, called E_p, which should satisfy the conditions of mechanical definition to calculate σ is given in Figures 2.b and d. The colour distribution is chosen so as to provide periodic boundary conditions in all directions; here a slice of one of the fluids is placed between two layers of the other fluid. One can expect that σ, obtained for both interfaces in this box, is the same as in the mechanical test in the limited box E_s.

Let the limited box E_s contain $N_x N_y N_z$ lattice nodes; its section perpendicular to the z-axis is $S = N_x N_y$; the channel is inclined with respect to the x-axis at

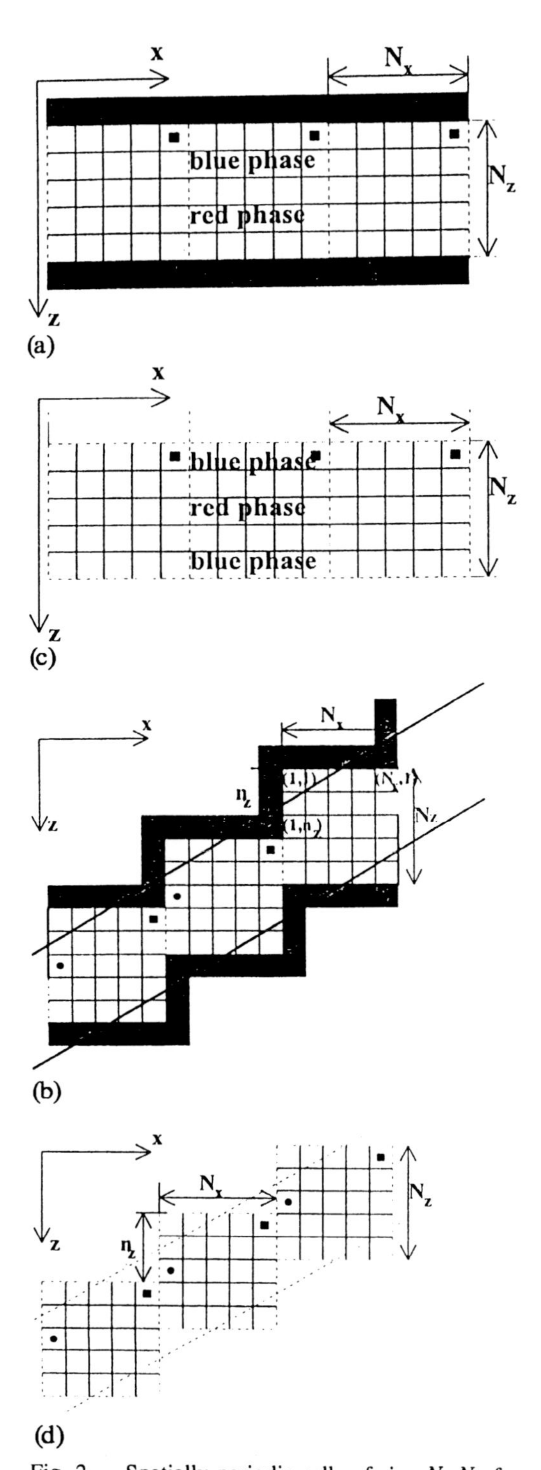

Fig. 2. Spatially periodic cells of size $N_x N_z$ for a 2D channel; $N_y = 1$. (a,c) The cell, called E_s, is limited by solid walls parallel to the interface. The solid walls correspond to the shaded areas. The periodic conditions are given by the broken lines. (b,d) The cell, named E_p, is periodic in all directions. The cells (a) and (b) are not inclined; (c) and (d) are inclined with an angle $\theta = \text{arctg } n_z/N_x$ with respect to the x-axis. In order to introduce this inclination, the usual periodic boundary conditions are modified in the cells (c) and (d); equivalent nodes are marked by the same symbols •, ▪. Phase distributions in (c) and (d) are given by the inclination of the phase distributions in (a) and (b), respectively.

an angle $0° \leqslant \theta \leqslant 45°$; this is provided by a modification of the usual boundary conditions (see Figure 2.c). Some boundary conditions have to be imposed to simulate no slip conditions at the solid walls. In this paper, the standard bounce-back reflections, discussed by Cornubert *et al.* (1991) and Ginzbourg and Adler (1994a), are used

$$\begin{aligned} N_{-i}(\mathbf{r}, t+1) &= N_i''(\mathbf{r}, t) \quad \text{if } \mathbf{r} + \mathbf{C}_i \in \text{ solid}, \\ \mathbf{C}_{-i} &= -\mathbf{C}_i. \end{aligned} \tag{3.4}$$

Thus, two approaches could be investigated to derive the value of σ from mechanical tests. First, one evaluates the entire sum Ω without any consideration about the form of each population N_i. Second, one analytically determines the solutions N_i to the Boltzmann equation supplemented with the perturbation (2.3) and with boundary conditions (3.4). The first approach has been already used (Gunstensen *et al.*, 1991; Gunstensen, 1992; Appert, 1993; Grunau, 1993) assuming that the following condition is satisfied by the solution

$$\sum_{\mathbf{r} \in E_s} N_i(\mathbf{r} + \mathbf{C}_i, t+1) = \sum_{\mathbf{r} \in E_s} N_i(\mathbf{r}, t), \quad \forall i = 1, \ldots, b_m. \tag{3.5}$$

Note that although relations (3.5) are obviously satisfied by the solutions in the period configuration E_p (Figures 2.b and d), there is no ground to postulate (3.5) in an arbitrarily inclined limited box if the boundary conditions (3.4) are imposed even for fluids of identical viscosity, since the real solutions in such configurations include fluctuations of density and anomalous currents. This will be further commented in this Section. Note that a new technique to derive the surface tension introduced in this paper can be applied even if the condition (3.5) is not satisfied.

3.2. DETERMINATION OF SURFACE TENSION BY THE MECHANICAL TEST

3.2.1. *Population Solutions in 2D Configurations of Limited Extent*

In order to estimate the surface tension (3.3), let us study the population solutions to the Boltzmann equation with perturbation in the mechanical tests. The perturbation (2.3.a) can be written in the form

$$\delta N_i(\mathbf{r}, t) = C^{\text{per.}}(\mathbf{r}, t) Q_{i\,nn}, \tag{3.6.a}$$

where (cf. (3.2d))

$$\begin{aligned} Q_{i\,nn} &= \left\{ \frac{(\mathbf{C}_i \cdot \mathbf{f})^2}{(\mathbf{f} \cdot \mathbf{f})^2} - \frac{1}{2} \right\} = C_{i\,n}^2 - \frac{1}{2} \\ &= \sum_{\alpha \in \{x,y,z\}} f_{e\alpha}^2 Q_{i\,\alpha\alpha} + \sum_{\alpha,\beta \in \{x,y,z\}} f_{e\alpha}^2 f^2 Q_{i\alpha\beta}, \quad Q_{0\,nn} = 0. \end{aligned} \tag{3.6.b}$$

The second order invariant tensors $Q_{i\ \alpha\beta}$, recalled in Appendix A, are eigenvectors of the collision matrix (Higuera *et al.*, 1989; Ginzbourg and Adler, 1994a) and they correspond to the eigenvalue λ_ψ which defines the bulk kinematic viscosity by (2.2.f).

The Boltzmann equation (2.1)–(2.5) without the external force (2.1.c) can be expressed with the help of (3.6) as

$$\begin{aligned} & N_i(\mathbf{r} + \mathbf{C}_i, t + 1) \\ & \quad = N_i(\mathbf{r}, t) + \sum_{j=0}^{b_m} A_{ij} N_j^{\text{neq.}}(\mathbf{r}, t) W_j + C^{\text{per.}}(\mathbf{r}, t) Q_{i\ nn}, \\ & \quad i = 0, \ldots, b_m. \end{aligned} \tag{3.7}$$

A trivial solution to the Boltzmann equation without the perturbation (3.6.b) in the configurations E_s and E_p (Figure. 2) is simply

$$N_i(\mathbf{r}) \equiv d = \text{const}, \quad \forall i = 0, \ldots, b_m, \tag{3.8}$$

where d is the cell density (cf. 2.1.b) defined as

$$d(\mathbf{r}, t) = \frac{\rho(\mathbf{r}, t)}{b}, \quad b = b_m + L. \tag{3.9}$$

However, any solution N_i can be expanded on the 25 basis vectors chosen among the eigenvectors of the FCHC collision matrix $\mathbf{A}$ with rest populations (Gunstensen, 1992; Ginzbourg and Adler, 1994). The complete set of the eigenvectors of $\mathbf{A}$ is described in Appendix A

$$\{\mathbf{e}^0,\ \mathbf{C}_\alpha,\ \mathbf{Q}_{\alpha\beta},\ \mathbf{T}_{\alpha\beta\beta},\ \mathbf{T}_{\alpha\alpha\beta\beta},\ \mathbf{e}^{\text{im}}\}, \quad \{\alpha, \beta\} \in \{x, y, z, w\}, \tag{3.10}$$

where w is the fourth component of the velocity in the FCHC lattice. For the 2D mechanical tests, the population solutions do not depend upon y and w; moreover, the velocity is equal to zero along these directions; consequently, the following expansion for these solutions in terms of the eigenvectors (3.10) is assumed

$$\begin{aligned} N_i(\mathbf{r}, t) \ = \ & d(\mathbf{r}, t) e_i^\circ + \sum_{\alpha=\{x,z\}} \Delta_\alpha(\mathbf{r}, t) C_{i\alpha} + \sum_{\alpha,\beta=\{x,z\}} \Delta_{\alpha\beta}(\mathbf{r}, t) Q_{i\ \alpha\beta} + \\ & + \sum_{\substack{\alpha \neq \beta \\ \alpha,\beta=\{x,z\}}} \Delta_{\alpha\beta\beta}(\mathbf{r}, t) T_{i\ \alpha\beta\beta} + \Delta_{xxzz}(\mathbf{r}, t) T_{i\ xxzz} + \\ & + \Delta^{im}(\mathbf{r}, t) e_i^{im}, \quad \forall i = 0, \ldots, b_m \end{aligned} \tag{3.11.a}$$

where $d(\mathbf{r}, t), \Delta_\alpha(\mathbf{r}, t), \Delta_{\alpha\beta}(\mathbf{r}, t), \Delta_{\alpha\beta\beta}(\mathbf{r}, t), \Delta_{\alpha\alpha\beta\beta}(\mathbf{r}, t)$ and $\Delta^{im}(\mathbf{r}, t)$ are the coefficients of the decomposition. Note that all the coefficients Δ can be uniquely derived from the set of population solutions N_i in the mechanical test. The coefficient $d(\mathbf{r}, t)$ corresponds to the cell density (3.9) because all other tensors in (3.11.a) conserve the local mass. The local velocity can be derived from (2.1.f) and (3.11.a)

$$u_\alpha(\mathbf{r}, t) = \frac{c^2 b_m}{D} \frac{\Delta_\alpha}{\rho(\mathbf{r}, t)}, \quad \alpha = \{x, z\}. \tag{3.11.b}$$

Note that Δ_α corresponds to the so-called anomalous currents; the other tensors vanish because of the contraction with the velocity $C_{i\alpha}$. We shall see in this Section that anomalous currents disappear *in the stationary solutions* for some particular choices of the eigenvalues found by solving the Boltzmann equation with perturbation (3.7) in the configurations with solid boundaries E_s; these solutions will be numerically confirmed in the next Section.

It can be easily shown that only second order eigenvectors $Q_{i\ \alpha\beta}$ can contribute to σ (see Appendix A). For example, for the 2D configurations displayed in Figure. 2 (cf. 3.6.b)

$$\sum_{i=1}^{24} Q_{i\ xx}\Phi_i = -8 \cos 2\theta, \qquad \sum_{i=1}^{24} Q_{i\ zz}\Phi_i = 8 \cos 2\theta,$$

$$\sum_{i=1}^{24} Q_{i\ xz}\Phi_i = -8 \sin 2\theta. \tag{3.12.a}$$

Moreover, this contribution does not depend upon the angle θ for the second-order perturbation tensors $Q_{i\ nn}$, defined by (3.6.b), since the sum $\sum_{i=1}^{b_m} Q_{i\ nn}\Phi_i$ is constant for any $Q_{i\ nn}$

$$\sum_{i=1}^{b_m} Q_{i\ nn}\Phi_i = 8. \tag{3.12.b}$$

So, in order to analytically derive the surface tension from the population solutions, one has to find the coefficients $\Delta_{\alpha\beta}(\mathbf{r}, t)$ associated with the second-order tensors in the solutions (3.11.a).

Let us now propose that the stationary solution is obtained in mechanical tests. In order to obtain such solutions, the perturbation has to be time independent

$$\mathbf{f}(\mathbf{r}, t) = \mathbf{f}(\mathbf{r}). \tag{3.13.a}$$

This condition is difficult to satisfy in inclined channels, where the phase distribution inside the interface layer can change between two or more iterations.

Thus, the standard perturbation (2.3.b) provides stationary solutions only if a time-independent colour distribution can be found by the model. This happens for example for noninclined interfaces such as the one given in Fig. 1.b when the red ($z \leqslant z^R$) and blue ($z \geqslant z^B$) fluids are always divided by a lattice link. If the colour distribution is periodic with a period $T = 2$, as in the case of the interface between the red and blue links (for example, the colour distribution given in Figure 1.a) without residual amounts of opposite colour at the nodes $z = z^R$ and $z = z^B$ (cf. 3.20d), time-independent total population solutions can be obtained with

$$C^{\text{per.}}(\mathbf{r},t) = A^{\text{per.}}\frac{\{|\mathbf{f}(\mathbf{r},t) + \mathbf{f}(\mathbf{r},t-1)|\}}{2} \stackrel{\text{def.}}{=} A^{\text{per.}}\langle|\mathbf{f}(\mathbf{r})|\rangle \qquad (3.13.\text{b})$$

However, in inclined channels, colour distributions may be more complicated than the ones displayed in Figure. 1 due to discretisation; consequently, time periodic patterns with periods larger than 2 can be obtained at the interfacial nodes. Actually, we cannot indicate a universal method to obtain stationary solutions for an arbitrary interfacial phase distribution. Nevertheless, it should be emphasised that all the results of this paper are obtained under the hypothesis that the solutions of the mechanical test are stationary and hence the solution (3.11.a) is assumed to be time-independent.

In time-independent solutions, the Boltzmann equation (3.7) for rest populations N_0 yields

$$\Delta^{im}(\mathbf{r}) = 0, \quad \forall\mathbf{r} \text{ if } \lambda_c \neq 0, \qquad (3.14.\text{a})$$

where λ_c is the eigenvalue corresponding to the eigenvector $\mathbf{e}^{im}$ related to the introduction of rest populations in the model (see A.4). In the following (cf. 3.15), the term with $\mathbf{e}^{im}$ is omitted in the solutions; in any case, we shall see that it does not influence surface tension. For symmetry reasons, 2D solutions for the configurations E_s and E_p should not depend upon y and w and upon the tangential direction when the interface is parallel with some lattice velocity; hence, the populations should also satisfy the following invariance conditions derived from the Boltzmann equation (3.7)

$$\sum_{j=0}^{24} A_{ij} N_j^{\text{neq.}}(\mathbf{r}) W_j = -C^{\text{per.}} Q_{i\ nn}, \qquad (3.14.\text{b})$$

for

$$i \in I_{0^\circ<\theta<45^\circ} \stackrel{\text{def.}}{=} \{i\colon C_{iy}C_{iw} = \pm 1\}, \qquad \text{for } 0^\circ < \theta < 45^\circ,$$

$$i \in I_{\theta=45^\circ} \stackrel{\text{def.}}{=} \{i\colon C_{iy}C_{iw} = \pm 1 \text{ or } C_{ix}C_{iz} = 1\}, \quad \text{for } \theta = 45^\circ,$$

$$i \in I_{\theta=0^\circ} \stackrel{\text{def.}}{=} \{i\colon C_{iz} = 0\}, \qquad \text{for } \theta = 0^\circ. \qquad (3.14.\text{c})$$

Consequently, the solutions (3.11.a) can be substituted into the Boltzmann equation (3.7) supplemented by the perturbation, the bounce-back boundary conditions (3.4) and the conditions (3.14.b)–(3.14.c) to extract the values of the coefficients Δ in the decomposition (3.11.a) in terms of the eigenvalues of the collision matrix. These coefficients also depend on the perturbation function $C^{\text{per.}}$

It can be easily shown that in non-inclined channels (Figures 2a, c) the general form of the *time-independent* solution to (3.7), (3.14) is

$$N_i(\mathbf{r}) = d(\mathbf{r}) - \frac{C^{\text{per.}}(\mathbf{r})}{\lambda_\psi(\mathbf{r})} Q_{i\ nn}, \quad i = 0, \ldots, 24 \quad \text{for } \theta = 0^\circ. \tag{3.15.a}$$

Therefore, the anomalous currents are absent in such solutions and the second-order terms in the decomposition (3.11.a) are proportional to the perturbation tensor $Q_{i\ nn}$.

The solutions with such properties can be obtained in inclined channels only if some special choice of the eigenvalues of the collision matrix is made. Assuming that a solution in the channel inclined at an angle $\theta = 45^\circ$ is time-independent, it is shown in Appendix B that the eigenvalue λ_1, associated with fourth-order eigenvectors (see Appendix A, (3.11.a)) has to be equal to zero in order to annihilate the anomalous currents and, simultaneously, to obtain second-order terms proportional to the perturbation tensor

$$N_i(\mathbf{r}) = d(\mathbf{r}) - \frac{C^{\text{per.}}(\mathbf{r})}{\lambda_\psi(\mathbf{r})} Q_{i\ nn} + \Delta_{xxzz}(\mathbf{r}) T_{i\ xxzz},$$

$$\lambda_1 \equiv 0, \quad \text{for } \theta = 45^\circ. \tag{3.15.b}$$

In fact, the choice $\lambda_1 \equiv 0$ means that the quantities associated with fourth-order invariant tensors ($\sum_i N_i T_{i\ xxzz}$) are additionally conserved during the collision.

In the arbitrary inclined channel ($0^\circ < \theta < 45^\circ$), in addition to the choice $\lambda_1 \equiv 0$, the eigenvalue λ_2 has to be equal to -2 in order to obtain the following time-independent solutions, as detailed in Appendix C

$$N_i(\mathbf{r}) = d(\mathbf{r}) - \frac{C^{\text{per.}}(\mathbf{r})}{\lambda_\psi(\mathbf{r})} Q_{i\ nn} + \Delta_{xzz}(\mathbf{r}) T_{i\ xzz} + \Delta_{zxx}(\mathbf{r}) T_{i\ zxx} +$$

$$+ \Delta_{xxzz}(\mathbf{r}) T_{i\ zzxx},$$

$$\lambda_1 \equiv 0, \qquad \lambda_2 \equiv -2 \quad \text{for } 0^\circ < \theta < 45^\circ. \tag{3.15.c}$$

In fact, the choice $\lambda_2 = -2$ leads to the conservation of staggered quantities (see Zanetti, 1991).

Note that the bounce-back conditions have been explicitly used to derive (3.15.b) and (3.15.c); however, the solutions for the eigenvalues do not depend on the interface position in the channel.

Some remarks are now in order. First, for $\theta = 0°, Q_{i\,nn}$ can be expressed as (cf. 3.6.b)

$$Q_{i\,nn}(\theta = 0°) = \begin{cases} -\frac{1}{2}, & i \in \mathrm{I}_{\theta=0°}, \\ \frac{1}{2}, & i \notin \mathrm{I}_{\theta=0°}. \end{cases} \tag{3.16.a}$$

Hence, the solution (3.15) for $\theta = 0°$ may be written as (cf. (3.14.c) for the definition of $I_{\theta=0°}$)

$$N_i(\mathbf{r}) = \begin{cases} d^{\text{real}} + \dfrac{C^{\text{per.}}(\mathbf{r})}{\lambda_\psi(\mathbf{r})}, & i \in \mathrm{I}_{\theta=0°}, \\ d^{\text{real}}, & i \notin \mathrm{I}_{\theta=0°}, \\ d^{\text{real}} + \dfrac{C^{\text{per.}}(\mathbf{r})}{2\lambda_\psi(\mathbf{r})}, & i = 0, \end{cases} \tag{3.16.b}$$

where

$$d^{\text{real}} = d(\mathbf{r}) - \frac{C^{\text{per.}}(\mathbf{r})}{2\lambda_\psi(\mathbf{r})}. \tag{3.16.c}$$

Note that the solutions N_i (3.15) for $\theta = 0°$ depend only on the z-coordinate of the vector $\mathbf{r}$

$$N_i(\mathbf{r}) = N_i(z), \quad \forall i = 0, \ldots, b_m. \tag{3.16.d}$$

It is easy to derive from (3.15) and (3.16) that the density d^{real} should be constant in order to satisfy the Boltzmann equation (3.7) and the bounce-back conditions (3.4). Actually, the solution coming from the node $\mathbf{r}$ into the neighbouring nodes may be obtained from the Boltzmann equation (3.7) and the solutions (3.15) for $\theta = 0°$ as

$$\begin{aligned} N_i(\mathbf{r} + \mathbf{C}_i) &= \left\{ d(\mathbf{r}) - \frac{C^{\text{per.}}(\mathbf{r})}{\lambda_\psi(\mathbf{r})} Q_{i\,nn} \right\} \overset{\text{collis}}{+} \{-C^{\text{per.}}(\mathbf{r}) Q_{i\,nn}\} \\ &\quad \overset{\text{perturb}}{+} C^{\text{per.}}(\mathbf{r}) Q_{i\,nn} \\ &= d(\mathbf{r} + \mathbf{C}_i) - \frac{C^{\text{per.}}(\mathbf{r} + \mathbf{C}_i)}{\lambda_\psi(\mathbf{r} + \mathbf{C}_i)} Q_{i\,nn} \end{aligned} \tag{3.16.e}$$

Then, from the definition (3.16.c) of d^{real} and from the relations (3.16.e) for sites with nonzero z-component of the velocity ($i \notin \mathrm{I}_{\theta=0°}$), one obtains

$$d^{\text{real}} = \text{const.} \tag{3.16.f}$$

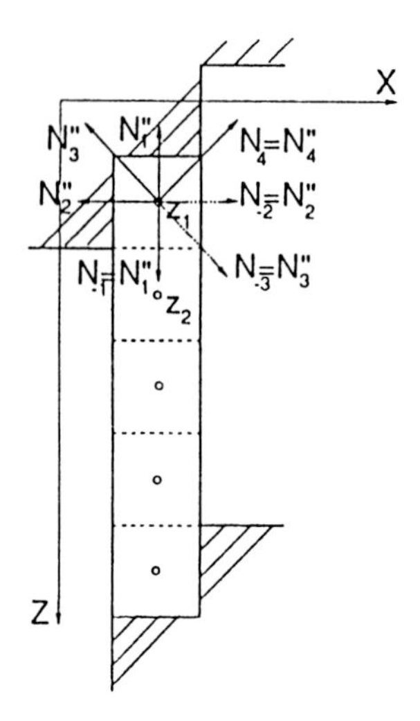

Fig. 3. Cell E_s of N_z lattice nodes ($N_x = 1, N_y = 1, n_z = 1$) for a channel inclined at $\theta = 45°$. The solid walls correspond to the hatched areas.

Consequently, the normal component p_n of the pressure tensor is constant according to the real physical pressure distribution in the mechanical test.

Thus, a constant density d^{real} moves between the nodes in the z-direction normal to the interface and density fluctuations for $\theta = 0°$ are due to the presence of the anisotropic terms $T_i^{\text{anisot.}}$ in the tangential populations $N_i (i \in \mathrm{I}_{\theta=0°})$ and the rest populations N_0

$$N_i(\mathbf{r}) = d^{\text{real}} + \frac{C^{\text{per.}}(\mathbf{r})}{\lambda_\psi(\mathbf{r})} T_i^{\text{anisot.}}, \quad \forall i = 0, \ldots, b_m, \tag{3.16.g}$$

where T_i^{anisot} should be different from zero only for $i \in \mathrm{I}_{\theta=0°}$ and for rest populations, as obtained from the comparison between (3.16.b) and (3.16.g); hence, for a FCHC lattice it can be written in the form

$$T_i^{\text{anisot.}} = \begin{cases} C_{i\,x}^2 C_{i\,w}^2 + C_{i\,x}^2 C_{i\,y}^2 + C_{i\,y}^2 C_{i\,w}^2, & i = 1, \ldots, 24, \\ \frac{1}{2}, & i = 0. \end{cases} \tag{3.16.h}$$

Note that thanks to the relation (3.16.f), the assumption (3.5) is satisfied by bounce-back or specular reflections in the non-inclined channels even if the viscosities are different; however, they are not satisfied in general when the additional terms associated with the eigenvectors $\mathbf{T}_{\alpha\beta\beta}$ and $\mathbf{T}_{\alpha\alpha\beta\beta}$ are present in the solution (cf. 3.15 for $\theta \neq 0°$). Indeed, a solution with $\Delta_{xxzz} \equiv 0$ as for $\theta = 0°$ cannot be obtained in an inclined configuration even for $\theta = 45°$ (Fig. 3). Actually, for $\theta = 45°$, the perturbation tensor $Q_{i\ nn}$ can be expressed as

$$Q_{i\ nn}(\theta = 45°) = \begin{cases} -\frac{1}{2}, & i \in \mathrm{I}_{\theta=45°}, \\ \frac{3}{2}, i & \text{such that } C_{i\ x}C_{i\ z} = -1, \\ 0, & \text{others:} \end{cases} \quad (3.17.a)$$

Hence, some constant density should move along horizontal and vertical directions if the solution (3.15.b) for $\theta = 45°$ verifies $\Delta_{xxzz} \equiv 0$. Then, the Boltzmann equation (3.7) along the direction normal to the interface should yield

$$\frac{C^{\text{per.}}(\mathbf{r})}{\lambda_\psi(\mathbf{r})} = \text{const.} \quad \forall \mathbf{r}. \quad (3.17.b)$$

But such a relation (3.17.b) is impossible since the perturbation function decreases with the distance from the interface. Consequently, the term Δ_{xxzz} has to be present in the solutions of this configuration. Nevertheless, it can be shown that at any node

$$\Delta_{xxzz} = \frac{3C^{\text{per.}}(\mathbf{r})}{\lambda_\psi(\mathbf{r})}. \quad (3.17.c)$$

Consequently, nontangential populations (3.15.b) ($i \notin \mathrm{I}_{\theta=45°}$) are constant and equal to d^{real} defined as in (3.16.c). This means that the normal component p_n of the pressure tensor is constant, similar to the solution (3.15.a) for a noninclined channel.

3.2.2. *Surface Tension*

The solutions in the general form (3.11.a) have been studied for a configuration arbitrarily inclined at an angle $0° \leqslant \theta \leqslant 45°$ with respect to the lattice lines. The results show that the relations (3.5) could be postulated without any additional assumption of the eigenvalues of the collision matrix and on the interface position only for a noninclined box with $\theta = 0°$. Even in this case, an approach used to derive σ from the models with equal viscosities seems to fail when the bulk viscosities are different. Nevertheless, because of the form of the solutions (3.15), one can proceed in a different manner to determine the surface tension.

This can be achieved by substituting (3.15) into the sum (3.3). A remarkable property of the solutions (3.15) is that a sum of the terms associated with the second order tensors $Q_{i\ \alpha\beta}$ is proportional to $Q_{i\ nn}(\theta)$ with an explicit pre-factor

$$\sum_{\alpha,\beta=\{x,z\}} \Delta_{\alpha\beta}(\mathbf{r}) Q_{i\ \alpha\beta} = -\frac{C^{\text{per.}}(\mathbf{r})}{\lambda_\psi(\mathbf{r})} Q_{i\ nn}. \quad (3.18)$$

The relation (3.18), which is valid when the solution is stationary and the eigenvalues satisfy the conditions given in (3.15), is crucial to deriving the value of σ in (3.3). Substitution of the obtained solutions (3.15) into (3.3) yields for any θ the following solution.

$$\sigma = -8 \int_{-\infty}^{+\infty} \frac{C^{\text{per.}}(n)}{\lambda_\psi(n)}\, dn. \tag{3.19.a}$$

Hence, in noninclined channels with a normal z, the surface tension can be approximated by a sum

$$\sigma = -8 \sum_i \frac{C^{\text{per.}}(z)}{\lambda_\psi(z)}. \tag{3.19.b}$$

In Model 1, the calculation of the sum in (3.19.b) for σ, when the form (3.13.b) is used for $C^{\text{per.}}(\mathbf{r})$, yields

$$\sigma^{(1)} = \sigma_0^{(1)} + \delta^{(1)}, \tag{3.20.a}$$

where the theoretical prediction for σ is

$$\begin{aligned} \sigma_0^{(1)} &= -96 d A^{\text{per.}} \left(\frac{24 + L^R}{\lambda_\psi^R} + \frac{24 + L^B}{\lambda_\psi^B} + \frac{1}{2}\left(\frac{1}{\lambda_\psi^B} - \frac{1}{\lambda_\psi^R}\right)(L^R - L^B) \right), \\ d &= d^{\text{real}}. \end{aligned} \tag{3.20.b}$$

For the noninclined interface and for the phase distribution displayed in Figure. 1.a, the error term $\delta^{(1)}$ may be found from the solution (3.16.g) in the form

$$\delta^{(1)} = c A^{\text{per.}^2} \left(\frac{1}{\lambda_\psi^B} - \frac{1}{\lambda_\psi^R} \right), \tag{3.20.c}$$

where c is some constant depending on the actual colour distribution in the nodes $\mathbf{z}^R$ and $\mathbf{z}^B$. For example, let us assume that in the result of the recolouring step (2.4a)–(2.4d), all the amount of red (blue) phase leaves the blue (red, respectively) boundary. Hence, the interfacial amounts ρ_I^B and ρ_I^R of each colour are related by

$$\rho_I^R(\mathbf{z}^R) + \rho_I^R(\mathbf{z}^B) = \rho(\mathbf{z}^R), \qquad \rho_I^B(\mathbf{z}^B) + \rho_I^B(\mathbf{z}^R) = \rho(\mathbf{z}^B) \tag{3.20.d}$$

Equivalently, the normal populations propagation between blue and red boundaries $\mathbf{z}^B$ and $\mathbf{z}^R$ have equal amounts of red and blue phases, equal to $d^{\text{real}}/2$. In that case, the value of the constant c in (3.20.c) is

$$c = -48 \left((24 + L^R) \frac{\langle |\mathbf{f}(\mathbf{z}^R)| \rangle}{2\lambda_\psi^R} - (24 + L^B) \frac{\langle |\mathbf{f}(\mathbf{z}^B)| \rangle}{2\lambda_\psi^B} \right). \tag{3.20.e}$$

In (3.20.b) and (3.20.e), d^{real}, given by (3.16.c), denotes the density in the non-perturbed region; the error term $\delta^{(1)}$ characterises the dependence of σ on the phase distribution at the interface.

Similar developments hold for Model 2 with the perturbation (2.3.b)

$$\sigma^{(2)} = \sigma_0^{(2)} + \delta^{(2)}, \tag{3.21.a}$$

$$\sigma_0^{(2)} = -48A^{\text{per.}}d\left(\frac{24+L^R}{\lambda_\psi^R} + \frac{24+L^B}{\lambda_\psi^B} + \frac{48+L^R+L^B}{\lambda_\psi^{\text{int}}}\right),$$

$$d = d^{\text{real}}, \tag{3.21.b}$$

with

$$\delta^{(2)} = \delta_1^{(2)} + \delta_2^{(2)},$$

$$\delta_1^{(2)} \cong -48A^{\text{per.}}(\rho_I^R - \rho_I^B)\left(\frac{1}{\lambda_\psi^B} - \frac{1}{\lambda_\psi^R}\right), \tag{3.21.c}$$

$$\delta_2^{(2)} = -48A^{\text{per.}^2}\left\{c^R\left(\frac{1}{\lambda_\psi^{\text{int.}}} - \frac{1}{\lambda_\psi^R}\right) + c^B\left(\frac{1}{\lambda_\psi^{\text{int.}}} - \frac{1}{\lambda_\psi^B}\right)\right\}.$$

where C^R and C^B are two constants which can be obtained as before using the real solutions (3.16.g) in noninclined channels. These two constants are

$$c^R = (24+L^R)\frac{|\mathbf{f}(\mathbf{z}^R)|}{2\lambda_\psi^R}, \qquad c^B = (24+L^B)\frac{|\mathbf{f}(\mathbf{z}^B)|}{2\lambda_\psi^B}. \tag{3.21.d}$$

Since anisotropic terms are present in the population solutions (cf. (3.16.g)–(3.16.h)), the usual macroscopic density (2.1.b) differs from the real simulated density ρ^{real} in the models with perturbation

$$\rho^{\text{real}} = (24+L)d^{\text{real}}. \tag{3.22}$$

These nonphysical fluctuations of density are proportional to $C^{\text{per.}}(\mathbf{r})/\lambda_\psi(\mathbf{r})$ and they induce errors $\delta^{(1)}$ and $\delta_2^{(2)}$ in real surface-tension measurements by comparison with the theoretical evaluations of the surface tensions (3.20.b) and (3.21.b) for fluids with different viscosities. Moreover, the real surface-tension value depends on the actual amount of each phase at the interface (cf. term $\delta_1^{(2)}$ in 3.21.c) when the perturbation is given in the form (2.3.b).

Let us finally go back to the case already investigated in the literature

$$L^R = L^B = L^{\text{int.}}, \qquad \lambda_\psi^R = \lambda_\psi^B = \lambda_\psi^{\text{int.}}. \tag{3.23}$$

Then, the theoretical predictions (3.20.b) and (3.21.b) coincide; if the mean cell density d_0 is substituted into these solutions and density fluctuations are neglected, they coincide with the solution of Gunstensen (1992). It is also interesting to notice that when the viscosities are equal, the corrective terms vanish. Moreover, the perturbation function $C^{\text{per.}}(\mathbf{r}, t)$ can be used in the form (cf. 2.3.b)

$$C^{\text{per.}}(\mathbf{r}, t) = A^{\text{per.}} \lambda_\psi(\mathbf{r}, t) \langle |\mathbf{f}(\mathbf{r}, t)| \rangle \tag{3.24}$$

the surface tension being derived as the series (cf. 3.19.b)

$$\sigma = -8 A^{\text{per.}} \sum_z \langle |\mathbf{f}(z)| \rangle. \tag{3.25}$$

The summation in (3.25) avoids the influence of the real-phase distribution at the interface on the surface tension because all the terms in this series disappear except for ones at the boundaries of the configuration ($z \to \pm\infty$).

4. Numerical

4.1. MECHANICAL TEST FOR SURFACE TENSION

4.1.1. *Model 1*

In order to verify the results (3.20)–(3.21) for σ, let us first consider mechanical tests in a non-inclined box limited by solid boundaries (Figure 2.a). In the simplest test for Model 1, the lattice is initialised by fluids with the same density per node ρ_0; since the number L of rest populations is the same for each phase, the density d_0 is equal to the mean cell density in the box

$$d_0 = \rho_0 / L. \tag{4.1}$$

The obtained phase distribution satisfies the conditions given in Figure 1.a; the position of the interface with respect to the box has not played any role in the previous calculations of σ. When the system reaches the steady state, the real density d^{real} (3.16.c) is derived from the nonperturbed nodes near the solid; to avoid any perturbation in the last lattice nodes, the solid has the colour and the density d^{real} of the attached fluid close to it. The difference between the mean density d_0 and d^{real} depends on the perturbation constant $A^{\text{per.}}$ and on the kinematic viscosities of the fluids. Roughly speaking, this difference increases with the perturbation parameter $A^{\text{per.}}/\lambda_\psi$ (cf. 3.16). This is demonstrated in Figure 4 by comparing the computations to the theoretical predictions (3.20.b) for $\sigma_0^{(1)}(d^{\text{real}})$ and $\sigma_0^{(1)}(d_0)$ as functions of the eigenvalues λ_ψ and of the perturbation constant $A^{\text{per.}}$. If the perturbation (2.3) is formally imposed in all the nodes of the channel rather than in the neighbourhood of the interface, the exact value of d^{real} slightly depends on the dimension of the box in the z-direction.

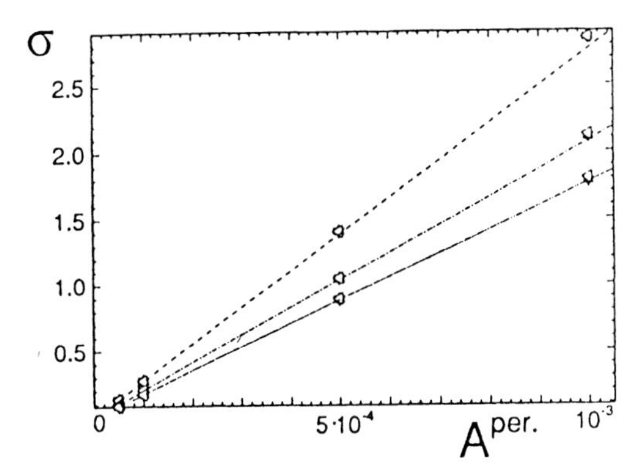

Fig. 4. Mechanical surface tension test by Model 1 in a non-inclined channel E_s. The calculated values $\sigma_{\text{def.}}$ (▼) and $\sigma_{\text{calcul.}}$ (▲) correspond to surface tension given by (3.3) and (3.19), respectively. The predictions $\sigma_0^{(1)}$ (3.20.b) are calculated for $d = d^{\text{real}}(\triangleleft)$ and for $d = d_0$ (straight lines). Data are for: $L^R = L^B = L^{\text{int.}} = 24, d_0 = 0.2$, $\lambda_\psi^R = -0.5$, $\lambda_\psi^B = -0.5$(- - - -); $\lambda_\psi^R = -0.5$, $\lambda_\psi^B = -1$ (-.-.-.-); $\lambda_\psi^R = -0.5$, $\lambda_\psi^B = -1.8$ (-..-..-..-). The dimensions of the channel are $1 \times 1 \times 19$ (in lattice units); the real density d^{real} is different for each measurement; the initial density per cell is equal to d_0.

At equilibrium, two values of σ are calculated; the first one σ_{def} is given by (3.3) as a direct consequence of the mechanical definition (3.1)–(3.2) and the second one $\sigma_{\text{calcul.}}$ from the theoretical solution (3.19). These values are seen to coincide for any perturbation constant $A^{\text{per.}}$. In agreement with theory, the numerical results show that this equality is valid for the stationary solution in any inclined box limited by solid boundaries if the eigenvalues of the collision matrix are chosen according to (3.15).

On the other hand, Figure 4 demonstrates the influence of the ratio between the kinematic bulk viscosities on the deviation of the real surface-tension coefficient (3.19) from the prediction $\sigma_0^{(1)}(d^{\text{real}})$ given by (3.20.b). For fluids with the same number of rest populations, this difference is proportional to $A^{\text{per.}^2}(1/\lambda_\psi^B - 1/\lambda_\psi^R)$ (see $\delta^{(1)}$ in 3.20.c). As confirmed in Fig. 4, when the viscosities are the same ($\lambda_\psi^R = \lambda_\psi^B$), the values of $\sigma_{\text{calcul.}}$ and $\sigma_0^{(1)}(d^{\text{real}})$ are equal (see $\lambda_\psi^R = \lambda_\psi^B = -0.5$) in contrast with the case $\lambda_\psi^R \neq \lambda_\psi^B$ where the viscosities are different (see $\lambda_\psi^R = -0.5$, $\lambda_\psi^B = -1.8$, for example). Hence, the theoretical predictions (3.20) are verified for Model 1.

4.1.2. *Model 2*

For Model 2, one can numerically observe the additional deviations of surface tensions measurements from the theoretical prediction (3.21.b) (see the error $\delta_1^{(2)}$ in 3.21.c) due to the presence of different amounts of each colour at the interface. We have seen that this term can be easily eliminated by the modification (3.25) of

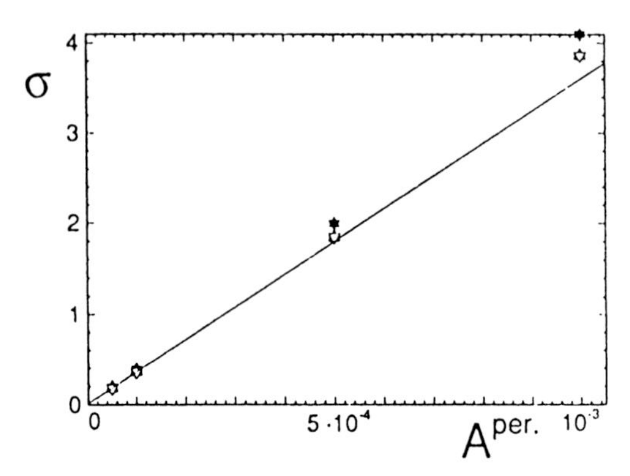

Fig. 5. Mechanical surface tension test by Model 2 in a non-inclined channel E_s. The measured values $\sigma_{\mathrm{def.}}$ and $\sigma_{\mathrm{calcul.}}$ correspond to surface-tension coefficients in the form (3.3) and (3.19), respectively; $\sigma_{\mathrm{def.}}$ ($\blacktriangledown$) and $\sigma_{\mathrm{calcul.}}$ ($\blacktriangle$) are obtained with the standard colour gradient (2.3.c); $\sigma_{\mathrm{def.}}$ (∇) and $\sigma_{\mathrm{calcul.}}$ (Δ) with the modified colour gradient (4.2). Solid line: theoretical solution (3.21.b) with $d = d_0$. The channel is $1 \times 1 \times 19$ (in lattice units); the initial cell density is different from d_0. Data are for: $L^R = 24$, $L^B = 36$, $L^{\mathrm{int.}} = 30$, $d_0 = 0.2476$, $\lambda_\psi^R = -1$, $\lambda_\psi^B = -0.5$.

the perturbation function; however, this term may be also annihilated if one takes a slightly modified colour gradient (cf. 2.3.c) for comparison)

$$\mathbf{f}^{\mathrm{mod.}}(\mathbf{r}, t) = \sum_{i=1}^{24} \tilde{f}_i \mathbf{C}_i,$$

$$\tilde{f}_i = \begin{cases} \rho_R(\mathbf{r} + \mathbf{C}_i, t) - \rho_B(\mathbf{r} + \mathbf{C}_i, t), & \text{if } \mathbf{r} + \mathbf{C}_i \notin \text{interface}, \\ 0, & \text{if } \mathbf{r} + \mathbf{C}_i \in \text{interface}. \end{cases} \tag{4.2}$$

This means that the modified colour gradient does not depend upon the interfacial amounts of each colour ρ_I^B and ρ_I^R. Then, the calculation of the sum in (3.19) shows that the terms $\delta_1^{(2)}$ in (3.21.c) as well as the term $\delta^{(1)}$ in (3.20.c) are formally annihilated in a noninclined box if the perturbation (2.3.b) corresponds to (4.2). This is confirmed in Figure 5.

The results in Figure 5 compare $\sigma_{\mathrm{def.}}$ (3.3) and $\sigma_{\mathrm{calcul.}}$ (3.19), derived from a mechanical test performed by Model 2 for both standard (2.3.c) and modified (4.2) colour gradients with the theoretical solution $\sigma_0^{(2)}$ (3.21.b). Here the numbers of rest populations are different in every phase; the lattice is initialised by fluids with different cell densities, so that the resulting interfacial position is found by the model according to the solution (3.16.b) of the Boltzmann equation (3.7) with perturbation. The resulting colour contributions ρ_I^B and ρ_I^R to the total interfacial density ρ are not equal ($\rho_I^B \neq \rho_I^R$). This yields an additional non-zero error term $\delta_1^{(2)}$ (3.21.c) of first order $O(A^{\mathrm{per.}})$ in comparison with the prediction (3.21.b) for surface tension in Model 2 if the standard colour gradient (2.3.c) is applied.

Moreover, σ is not a perfectly linear function of $A^{\text{per.}}$ for large values of $^{\text{per.}}$; in these conditions, the term $\delta_2^{(2)}$ of order $O(A^{\text{per.}^2})$ becomes significant. However, when a colour gradient of the form (4.2) is imposed, the total difference $\delta_1^{(2)} + \delta_2^{(2)}$ with the solution (3.21.b) decreases according to the prescribed values.

As in Model 1, the numerical values of $\sigma_{\text{def.}}$ (3.3) and $\sigma_{\text{calcul.}}$ (3.19) are equal in any limited configuration when the eigenvalues of the collision matrices satisfy the conditions (3.15) and when the solution is time-independent. Hence, the formula (3.19) for surface tension is numerically verified by mechanical tests in both models.

4.2. LAPLACE LAW

In order to verify the predictions (3.20.b) and (3.21.b) at curved interfaces, the lattice was initialised with a bubble of radius R in a box filled by a different fluid. The system is spatially periodic in all directions; different bulk kinematic viscosities are chosen. Then, according to the 2D Laplace law, the pressure difference between the bubble and the external fluid is inversely proportional to the bubble radius.

$$\Delta p = p_{\text{in}} - p_{\text{out}} = \frac{\sigma}{R}, \tag{4.3}$$

where σ is the same as in the mechanical definition (3.1).

In lattice gas methods, the pressure is related to the macroscopic density ρ by means of the constant C_s^2 (Frisch *et al.*, 1986; D'Humières and Lallemand, 1987), which is the sound velocity of the model

$$p = C_s^2 \rho, \qquad C_s^2 = \frac{c^2 b_m}{D(24 + L)}. \tag{4.4}$$

Note that in the simulations with different numbers of rest populations L in the two bulks or at the interface (cf. 2.2), the values of C_s^2 are different.

In order to eliminate the nonphysical density fluctuations due to the perturbation near the interface, the pressures p_{in} and p_{out} are measured at some distance from the interface

$$p_{\text{in}} = \langle p_{\mathbf{r}^2 \leqslant 0.5R^2}(\mathbf{r}) \rangle, \quad p_{\text{out}} = \langle p_{\mathbf{r}^2 \geqslant 1.7R^2}(\mathbf{r}) \rangle. \tag{4.5}$$

This is similar to the 'bubble test' of Gunstensen *et al.* (1991). It is assumed that the origin is at the bubble centre. In the region where the volume averaging is performed, the nodes are occupied by only one colour.

Then, the pressure difference (4.3), which is calculated numerically by means of the relations (4.4), is compared with the theoretical prediction (4.3) with σ given by (3.21.b) (see solid line in Figure 6) for the two colour gradients (2.3.c) and (4.2). Of course, the agreement between theory and numerical simulations is worse in

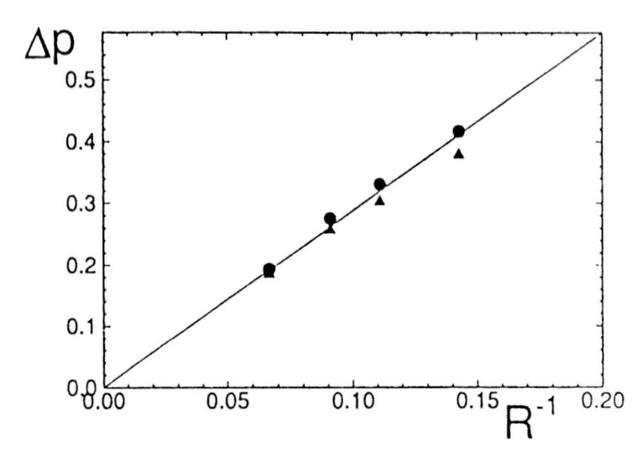

Fig. 6. Comparison between two numerical solutions computed by Model 2 with the Laplace law. Solid line: the Laplace law with the theoretical solution (3.21.b) for $d = d_0$; Δp is calculated by (4.5) from the simulations with the standard colour gradient (2.3.c) (•) and with a modified one (4.2) (▲). The periodic box E_p has a size $49 \times 1 \times 49$ (in lattice units) for a bubble radius $R = 15$ or $35 \times 1 \times 35$ for $R = 7, 9, 11$, repectively. Data are for: $A^{\text{per.}} = 5 \times 10^{-4}$, $L^R = L^B = L^{\text{int.}} = 24$, $\nu^R = 0.5$, $\nu^B = 1.5$, $\nu^{\text{int.}} = 0.75$, $d_0 = 0.2$.

general for small bubbles because of discretization. However, one also sees that Model 2 with a standard colour gradient has a better interfacial behaviour than a modified one as was expected on physical grounds. Hence, in spite of a formal improvement of the surface-tension coefficients (3.20.a) and (3.21.a), the modified colour gradient (4.2) cannot be recommended for simulations in real geometries.

4.3. ANOMALOUS CURRENTS

Anomalous currents appear in ILB (Gunstensen, 1992) and *liquid–gas* (Appert, 1993) near the interface if the normal to the interface is not parallel to the x-, y- or z-axes; these currents can be observed even in the case of non-inclined interfaces (Figures 2.a and b) in non-stationary solutions. For example, for interfaces inclined with an angle $\theta = 45°$, given in Figure 2 when $N_x = n_z$, or equivalently, in Figure 3, these currents are perpendicular to the interface. None of the many previous attempts to eliminate these spurious currents have succeeded.

In order to study the problem of appearance of these anomalous currents as well as in order to derive surface-tension coefficients for Models 1 and 2 for different bulk viscosities, we have studied the solution of the Boltzmann equation with the perturbation (3.7) in 2D boxes arbitrarily oriented with respect to the lattice axes. The results are quite obvious; anomalous currents exist in inclined configurations (Figures 2.b or d) because the solutions (3.11.a) with $\Delta_\alpha \neq 0, (\alpha = \{x, z\})$, satisfy the Boltzmann equation with the perturbation (3.7) and bounce-back boundary conditions (3.4) or periodic conditions.

The solutions for populations in a limited box oriented with an angle $\theta = 45°$ or $0° < \theta < 45°$ are given in Appendices B and C, respectively. These solutions can be used to eliminate the spurious currents by a special choice of the eigenvalues of the collision matrix. The following results have been obtained. The eigenvalue λ_1 of the collision matrix corresponding to the 4-order invariant lattice tensors $T_{i\ \alpha\alpha\beta\beta}$ (see Appendix A) is responsible for the appearance of spurious currents in time-independent solutions when $\theta = 45°$. The choice of $\lambda_1 = 0$ eliminates the currents in both boxes E_s and E_p as shown in Figure 7. In Figure 7.a, the eigenvalue λ_1 is taken equal to zero; consequently, the anomalous currents $(u_x,\ 0,\ -u_z)$, perpendicular to the interface, vanish with time. Currents appearing in the solutions for this configuration are shown in Figure 7.b as functions of λ_1. In the solutions with $\lambda_1 \neq -1$, the velocity converges in a monotone manner toward zero when λ_1 tends toward zero; however, it is remarkable that for $\lambda_1 = -1$ the numerical solution is not stable; consequently, the solutions given in the Appendix B which lead to the solution (3.15) when λ_1 tends toward zero are not valid anymore.

The anomalous currents can be eliminated for an arbitrarily inclined box limited by solid boundaries by a simultaneous use of the eigenvalues

$$\lambda_1 = 0, \quad \lambda_2 = -2, \tag{4.6}$$

λ_2 corresponds to the 3rd-order invariant lattice tensor $T_{i\ \alpha\alpha\beta}$ (see Somers and Rem, 1991 for FHP; Ginzbourg and Adler, 1994a, for FCHC, Appendix A). Although the eigenvalues λ_2 and λ_1 in (4.6) are at the limits of the interval $]-2,\ 0]$ established for eigenvalues of the collision matrix by a linear stability analysis (Higuera and Jimenez, 1989), the numerical solutions in mechanical tests converge to some reasonable time-independent or periodical solutions when these eigenvalues are taken. Note that time-independent solutions are not necessarily found in inclined configurations even if the eigenvalues λ_1 and λ_2 satisfy the conditions (4.6). Numerical experiments show that such solutions are easier to observe in boxes which are small in the z-direction, probably when the influence of the boundary conditions and of the initial phase distribution are more important. This is demonstrated in Figure 8. An inclined box E_s with $N_x = 2$, $n_z = 1$ ($\mathrm{tg}\,\theta \approx 1/2$) is used; it has only 5 lattice nodes in the z-direction ($N_z = 5$). Although the solutions with $\lambda_1 \neq 0$ (see Figure 8.a) or with $\lambda_2 \neq -2$ (Figure 8.b) depend upon time, they converge to the predicted solution (3.15) in the limit $\lambda_1 \to 0$ (Figure 8.a) or $\lambda_2 \to -2$ (Figure 8.b), respectively. The calculations are made in the same manner as above, using Model 2 with perturbation (2.3.b) in each lattice node.

When $\theta = 45°$ and $\lambda_1 = 0$, it is shown in Appendix B that the solution to the Boltzmann equation (3.6.b) in a periodic configuration E_p not limited by solid boundaries (Figure 2.d) corresponds to a constant velocity. Numerical calculations confirm that the solution with zero velocity is indeed obtained by the models (2.1)–(2.5) when the initial distribution of population has zero total momentum.

However, for an inclined periodic system of fluids with $0° < \theta < 45°$, the solution (3.15) without anomalous currents is not valid anymore even if the eigenvalues (4.6) are imposed. Numerical calculations confirm that the anomalous currents in a periodic box do not disappear even in the time-independent solutions in contrast with a limited box. The occurrence of anomalous currents is explained (Gunstensen,

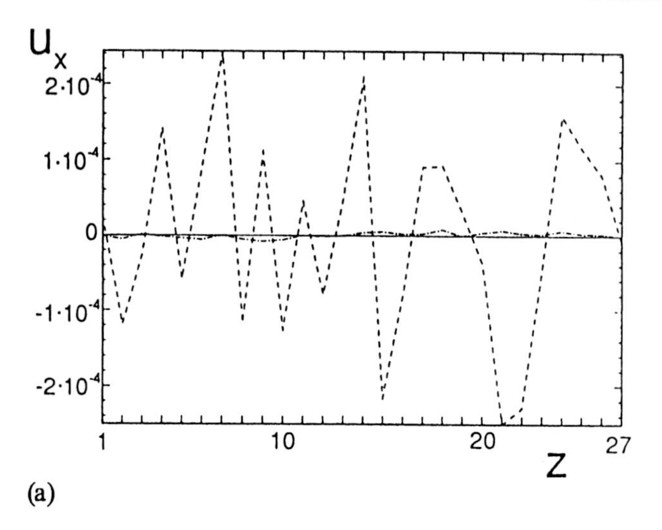

(a)

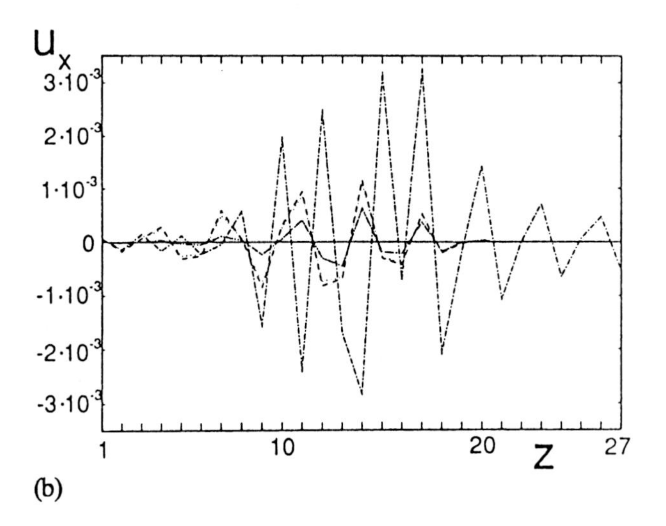

(b)

Fig. 7. Anomalous currents $\mathbf{u} = (u_x, 0, -u_x)$ obtained by Model 2 with perturbation (2.3.b) in the channel E_s inclined at $\theta = 45°$. The channel size is $1 \times 1 \times 27$ (in lattice units); the red fluid is at the bottom. Data are for: $A^{\text{per.}} = 5 \times 10^{-4}$, $L^R = L^B = L^{\text{int.}} = 0$, $\nu^R = 0.01852$, $\nu^B = 1.5$, $\nu_c^{\text{int.}} = 0.03659$, $d_0 = 0.3$. The dimensionless time T is given in number of iterations. (a) $\lambda_I = 0$: $T = 1000$(- - - -); $T = 2000$(-.-.-.-); $T = 4000$(——). (b) The currents are displayed for $T = 4000$ for various values of λ_I; $\lambda_I = 0$(— — —— ——); $\lambda_I = -0.5$(-..-..-..); $\lambda_I = -1$(-.-.-.); $\lambda_I = -1.5$(.......); $\lambda_I = -1.75$(- - - - - -) ; the final solutions are time-independent except for $\lambda_I = -1$.

1992; Appert, 1993) by the presence of density gradients due to the discretisation of the interface. However, the present analysis shows that in limited configurations, density fluctuations between neighbouring nodes could exist (cf. (3.16)) while anomalous currents are suppressed by the choice of the eigenvalues. Nevertheless, the anomalous currents cannot be always eliminated or reduced by the same choice of the eigenvalues.

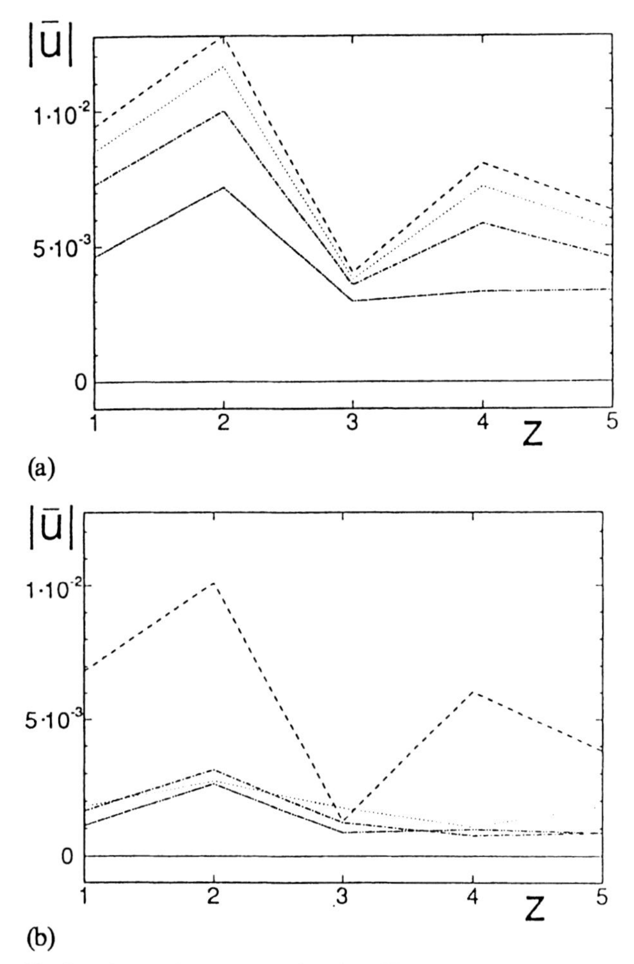

Fig. 8. Anomalous currents in a box E_s inclined at an angle $\theta \approx \text{arctg}\,(1/2)$ with respect to the x-axis ($N_x = 2,\ n_z = 1$); the cell has only 5 units in the z-direction ($N_z = 5$) with the red fluid at the bottom. The displayed velocity $\bar{\mathbf{u}}$ is calculated as the arithmetic mean of the currents in the two columns of the cell. Data are for: $A^{\text{per.}} = 5 \times 10^{-4}$, $L^R = L^B = L^{\text{int.}} = 0$, $d_0 = 0.3$, $\nu^R = 0.05(5)$, $\nu^B = 0.1(1)$, $\nu^{\text{int.}} = 0.0357$; (a) $\lambda_1 \to 0$, $\lambda_2 = -2$: $\lambda_1 = 0$(——); $\lambda_1 = -0.5$(-..-..-..); $\lambda_1 = -1$(-.-.-); $\lambda_1 = -1.5$(......); $\lambda_1 = -1.75$(- - - - - -). (b) $\lambda_2 \to -2$, $\lambda_1 = 0$: $\lambda_2 = -2$(——); $\lambda_2 = -1.5$ (-..-..-..); $\lambda_2 = -1$(-.-.-); $\lambda_2 = -0.5$(......); $\lambda_2 = 0$(- - - - - -).

4.4. DENSITY FLUCTUATIONS

In order to exhibit nonphysical fluctuations of density, we compare in Figure 9 two numerical solutions obtained by Model 2 with the perturbation (2.3) and a theoretical solution for a fully-established plane Poiseuille flow in a non-inclined channel

$$
\begin{aligned}
-F_x^B &= \nu^B \frac{\partial^2 u_x^B}{\partial z^2}, \quad z_0 \leqslant z \leqslant H, \\
-F_x^R &= \nu^R \frac{\partial^2 u_x^R}{\partial z^2}, \quad -H \leqslant z \leqslant z_0.
\end{aligned} \tag{4.7}
$$

Note that Model 2 without perturbation with the special interfacial collision matrix (see Table I) yields the exact solutions of (4.7) at all lattice nodes, except at the interface $z = z_0$ (Ginzbourg and Adler, 1994b).

Let us consider the numerical solution of the Boltzmann equation (3.6) supplemented by an interfacial perturbation in the non-inclined cell E_s. First, let us follow the standard scheme (2.1)–(2.5) when one adds an external force $\delta N_i^F(z)$ given by (2.1.c); $\delta N_i^F(z)$ is proportional to the density $\rho(z)$ defined as in (2.1.b). The results of Model 2 with the perturbation (2.3.b) are shown in Figure 9. The macroscopic density, which is derived in the usual way (2.1.b), includes non-physical fluctuations (cf. 3.16) due to the perturbation which are significant near the interface at nodes $z = z_0 \pm 1$. This explains the difference with the theoretical solution in these nodes in contrast with the models without perturbation.

The second simulation in Figure 9 corresponds to the calculation of the macroscopic velocity (2.1.f) by Model 2 with perturbation, but the density (2.1.b) in the relations (2.1.d) and (2.1.f) is replaced by the density ρ^{real} given by (3.22) and d^{real} is derived from the solutions in the nonperturbed nodes (cf. (3.16.g)). The solution is the same as for Model 2 without perturbation. This means that the populations in the fully established Poiseuille solution obtained with such a modified model are of the form

$$
\begin{aligned}
N_i(\mathbf{r}) \;=\; & \rho^{\text{real}} \left[b^{-1} + \frac{D}{c^2 b_m} \left(u_x C_{i\,x} - \frac{1}{2} F_x C_{ix} + \right.\right. \\
& \left.\left. + \frac{1}{\lambda_\psi} \frac{\partial u_x}{\partial z} C_{ix} C_{iz} + \frac{\nu}{\lambda_2} \frac{\partial^2 u_x}{\partial z^2} T_{i\,xzz} \right) \right] + \\
& + \frac{C^{\text{per.}}(\mathbf{r})}{\lambda_\psi(\mathbf{r})} T_i^{\text{anisot.}} + O(u^2)
\end{aligned} \tag{4.8}
$$

where the expression within the brackets corresponds to the non-perturbed LB solution in which the usual density (2.1.b) is replaced by $\rho^{\text{real.}}$, the reader is referred to a general form of populations, developed up to first order by Frisch *et al.* (1987), d'Humières and Lallemand (1987), and up to second order by Ginzbourg and Adler (1994a).

However in real simulations, the macroscopic density d^{real} cannot be derived from the population solutions by lattice gas methods. Hence, the macroscopic density field ρ and the velocity field $\mathbf{u}$ (2.1.f) are determined from (2.1.b) and

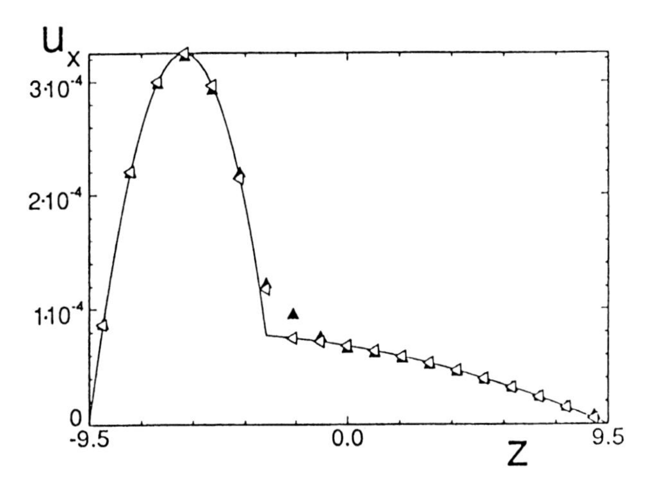

Fig. 9. Comparison between the two numerical solutions computed by Model 2 with the perturbation (2.3.b) and the analytical two-phase Poiseuille flow. Solid line: analytical solution; (▲) corresponds to the standard solution by Model 2 with perturbation; (Δ) corresponds to the solution by Model 2 with perturbation and with the density correction ($\rho = \rho^{\text{real}}$); the solution (Δ) is the same one as with Model 2 without perturbation. Data are for: $F_x^R = F_x^B = 10^{-6}$, $z_0 = -3$, $H = 9.5$, $A^{\text{per.}} = 5 \times 10^{-4}$, $L^R = L^B = L^{\text{int.}} = 24$, $\nu^R = 0.01852$, $\nu^B = 1.5$, $\nu^{\text{int.}} = 0.03569$.

(2.1.f) with the error of first-order $\mathrm{O}(A^{\text{per.}})$ near the interface where the Boltzmann equation (3.7) supplemented by the perturbation is used.

5. Conclusion

We have analytically studied the real solutions generated by ILB models in mechanical tests for fluids with different viscosities and numbers of rest populations. In general, such solutions contain nonphysical fluctuations of density and anomalous velocities; the last ones become particularly noticeable in inclined channels. In spite of the fact that solutions for the eigenvalues of the collision operator which suppress the anomalous currents in boxes of limited extent, have been found, we have to conclude that the density fluctuations and the anomalous currents are 'intrinsic' properties of ILB caused by the introduction of the perturbation into the single-phase lattice-Boltzmann equation. This means that the standard lattice gas procedure to derive the macroscopic fields from the population solutions, yields an error of order $\mathrm{O}(C^{\text{per.}}(\mathbf{r})/\lambda_\psi(\mathbf{r}))$. Consequently, quantitative numerical interfacial experiments seem to be seriously restricted by this lack of precision.

Although the method used in this paper to define surface tension with the help of the general form of solutions in appropriate configurations seems to be universal, it should be noticed that the solutions for the eigenvalues are closely related to the ILB equation on the FCHC lattice. Hence, they could be automatically applied neither to the 2D FHP ILB models, nor to the liquid–gas Boltzmann models. Moreover, these solutions cannot be recommended now for regular applications to the FCHC lattice since they are found to be at the limits of the convergence intervals.

The simplicity with which surface tension can be predicted from the mechanical definition, at least to leading order in terms of the perturbation constant $A^{\text{per.}}$, is the most attractive feature of ILB multiphase models. For sake of completeness, it would be interesting to derive σ analytically from the Laplace law and to compare it with the predictions made from the mechanical definition. Probably, by the same approach as here, the general population solutions in the 'bubble test' have to be first introduced using the symmetry properties of 'bubble test' configurations and of the FCHC lattice. Then the scheme of Somers and Rem (1991), developed for their FHP LG two-phase model, could be tentatively tried. We also expect that future quantitative comparisons of numerical solutions by multiphase ILB models (or by similar ones) with real interfacial experiments or with multiphase solutions, obtained by different methods, would verfiy the validity of derived predictions and evaluate the sphere of possible applications of these models.

Appendix A

A.1. EIGENVALUES AND EIGENVECTORS OF THE COLLISON MATRIX

The coefficients of the collision matrix **A** in FCHC Boltzmann model are defined by six values a_0, a_{60}, a_{90}, a_{120}, a_{180}, b_0 (see Higuera *et al.*, 1989; Gunstensen, 1992; Ginzbourg and Adler, 1994a). The matrix **A** can be completely defined by the choice of its eigenvalues. The eigenvalues and eigenvectors of **A** can be given as following

- an invariant tensor of zero order $\mathbf{e}^0$: $e_0^0 = L$, $e_i^0 = 1$, $i = 1, \ldots, 24$ associated with the eigenvalue zero of multiplicity 1; (A.1)
- invariant tensors of first order $C_{i\alpha}$, $\alpha = 1, \ldots, D$ associated with the eigenvalue zero of multiplicity D; (A.2)
- invariant tensors of second order $Q_{i\alpha\beta}$, $\alpha = 1, \ldots, D$, $\beta = 1, \ldots, D$

$$Q_{i\,\alpha\beta} = C_{i\,\alpha} C_{i\,\beta} - \delta_{\alpha\beta} c^2/D, \quad i = 1, \ldots, b_m; \quad Q_{0\,\alpha\beta} = 0,$$

associated with the eigenvalue λ_ψ of multiplicity 9

$$\lambda_\psi = a_0 - 2a_{90} + a_{180},$$

these tensors conserve mass and momentum

$$\sum_{i=1}^{24} Q_{i\,\alpha\beta} = 0; \quad \sum_{i=1}^{24} Q_{i\,\alpha\beta} C_{i\,\gamma} = 0, \quad \forall\gamma = 1, \ldots, D; \tag{A.3}$$

- invariant tensors of third order $T_{i\alpha\alpha\beta}$

$$T_{i\ \alpha\alpha\beta} = C_{i\ \beta} - 3C_{i\ \alpha}^2 C_{i\ \beta}, \quad i = 1, \ldots, b_m; \quad T_{0\ \alpha\alpha\beta} = 0$$

associated with the eigenvalue λ_2 of multiplicity 8

$$\lambda_2 = \tfrac{3}{2}(a_0 - a_{180}); \tag{A.4}$$

– invariant tensors of 4 – order $T_{i\ \alpha\alpha\beta\beta}$

$$T_{i\ \alpha\alpha\beta\beta} = C_{i\ \alpha}^2 C_{i\ \beta}^2 - \tfrac{1}{2}Q_{i\ \alpha\alpha} - \tfrac{1}{2}Q_{i\ \beta\beta} - \tfrac{1}{6}, \quad \forall \alpha \neq \beta; \quad T_{0\ \alpha\alpha\beta\beta} \equiv 0$$

associated with the eigenvalue λ_1 of multiplicity 2

$$\lambda_1 = \tfrac{3}{2}(a_0 + 6a_{90} + a_{180}) + 12\frac{b_0 L}{b_m}, \tag{A.5}$$

the eigenvectors $T_{i\ \alpha\alpha\beta}$ (A.4) and $T_{i\ \alpha\alpha\beta\beta}$ (A.5) conserve mass and momentum in the same manner as in (A.3);

– the zero-order eigenvector $\mathbf{e}^{\mathrm{im}}$ corresponds to the expansion of the FCHC collision matrix when L rest populations are included into the model (see Gunstensen, 1992; Ginzbourg and Adler, 1994a)

$$e_0^{i\ m} = -\frac{b_m}{L}, \quad e_i^{i\ m} = 1, \quad i = 1, \ldots, 24.$$

The corresponding eigenvalue of multiplicity 1 is

$$\lambda_c = -b_0 \left(\frac{b_m}{L} + L \right) \tag{A.6}$$

A.2. CONTRIBUTIONS OF THE EIGENVECTORS OF THE FCHC COLLISION MATRIX INTO THE SURFACE-TENSION COEFFICIENT OF THE ILB MODELS

In order to calculate the sum (3.3), let us define

$$\mathbf{A}^{(k)} = \sum_{i=1}^{24} \mathbf{T}_i^{(k)} \Phi_i \tag{A.7}$$

It should be evaluated for any eigenvector $\mathbf{T}_i^{(k)}$ of order k ($k = 0, \ldots, 4$) and for any vector Φ which may be introduced in (3.3)

$$\Phi_i = Q_{i\ nn} - Q_{i\ tt} = \sum_{\alpha \in \{x,y,z\}} b^{(\alpha)} Q_{i\ \alpha\alpha} + \sum_{\alpha \neq \beta} b^{(\alpha\beta)} Q_{i\ \alpha\beta} \tag{A.8.a}$$

The coefficients $b^{(\alpha)}$ and $b^{(\alpha\beta)}(\alpha, \beta = \{x, y, z\})$ are defined as (cf. also 3.2)

$$b^{(\alpha)} = (f_{e\alpha})^2 - (f_{e\alpha}^{\perp})^2, \quad b^{(\alpha\beta)} = f_{e\alpha} f_{e\beta} - f_{e\alpha}^{\perp} f_{e\beta}^{\perp} \tag{A.8.b}$$

$$\sum_{\gamma\in\{x,y,z\}} b^{(\gamma)} = 0 \tag{A.8.c}$$

where $\mathbf{f}_e$ is an arbitrary unit vector

$$\sum_{\alpha\in\{x,y,z\}} (f_{e\,\alpha})^2 = 1, \quad \sum_{\alpha\in\{x,y,z\}} (f_{e\,\alpha}^{\perp})^2 = 1, \quad \sum_{\alpha\in\{x,y,z\}} f_{e\,\alpha} f_{e\,\alpha}^{\perp} = 0. \tag{A.8.d}$$

So, the vector Φ_i is completely decomposed by the second-order eigenvectors $Q_{i\,\alpha\beta}$. It can be shown that it is orthogonal to any eigenvector of zero, first, third and fourth order (see (A.1), (A.6), (A.4) and (A.5), respectively). Thus, only second-order eigenvectors can contribute to the value of the surface-tension coefficient (3.3)

$$\begin{aligned} A_{\alpha\alpha}^{(2)} &= \sum_{i=1}^{24} Q_{i\,\alpha\alpha}\Phi_i = \sum_{\gamma\in\{x,y,z\}} b^{(\gamma)} \sum_{i=1}^{24} c_{i\,\alpha}^2 c_{i\,\gamma}^2 + \sum_{\gamma\neq\delta} b^{(\gamma\delta)} \sum_{i=1}^{24} c_{i\,\alpha}^2 c_{i\,\gamma} c_{i\,\delta} \\ &= \sum_{\gamma} b^{(\gamma}(8\delta_{\alpha\gamma} + 4) = 8 \sum_{\gamma} b^{(\gamma)} \delta_{\alpha\gamma}, \end{aligned} \tag{A.9.a}$$

$$\begin{aligned} A_{\alpha\beta}^{(2)} &= \sum_{i=1}^{24} Q_{i\,\alpha\beta}\Phi_i \\ &= \sum_{\gamma\in\{x,y,z\}} b^{(\gamma)} \sum_{i=1}^{24} c_{i\,\alpha} c_{i\,\beta} c_{i\,\gamma}^2 \\ &\quad + \sum_{\gamma\neq\delta} b^{(\gamma\delta)} \sum_{i=1}^{24} c_{i\,\alpha} c_{i\,\beta} c_{i\,\gamma} c_{i\,\delta} \\ &\overset{\text{FCHC}}{=} 4 \sum_{\gamma\neq\delta} b^{(\gamma\delta)} (\delta_{\alpha\beta}\delta_{\gamma\delta} + \delta_{\alpha\gamma}\delta_{\beta\delta} + \delta_{\alpha\delta}\delta_{\beta\gamma}) \\ &= 4 \sum_{\gamma\neq\delta} b^{(\gamma\delta)} (\delta_{\alpha\gamma}\delta_{\beta\delta} + \delta_{\alpha\delta}\delta_{\beta\gamma}), \quad \forall\alpha \neq \beta \end{aligned} \tag{A.9.b}$$

The relation (3.7.d) can be easily derived from (A.9) if $Q_{i\ nn}$ is decomposed into two parts

$$Q_{i\ nn} = \sum_{\alpha\in\{x,y,z\}} (f_{e\ \alpha})^2 Q_{i\ \alpha\alpha} + \sum_{\alpha\neq\beta} f_{e\ \alpha} f_{e\ \beta} Q_{i\ \alpha\beta} \tag{A.10}$$

Then, using (A.9a), (A.8) and (A.9b)

$$\sum_{i=1}^{24} Q_{i\ nn}\Phi_i = S_1 + S_2, \tag{A.11}$$

where

$$\begin{aligned} S_1 &= \sum_{\gamma\in\{x,y,z\}} (f_{e\ \gamma})^2 \sum_{i=1}^{24} Q_{i\ \gamma\gamma}\Phi_i = 12 \sum_{\gamma\in\{x,y,z\}} (f_{e\ \alpha})^4 - \\ &\quad -12 \sum_{\gamma\in\{x,y,z\}} (f_{e\ \gamma})^2 (f^{\perp}_{e\ \gamma})^2 + 4 \sum_{\gamma\in\{x,y,z\}} (f_{e\ \gamma})^2 \sum_{\delta\neq\gamma} [(f_{e\ \delta})^2 - (f^{\perp}_{e\ \delta})^2] \\ &= 8 \sum_{\gamma\in\{x,y,z\}} (f_{e\ \gamma})^4 - 8 \sum_{\gamma\in\{x,y,z\}} (f_{e\ \gamma})^2 (f^{\perp}_{e\ \gamma})^2, \\ S_2 &= \sum_{\gamma\neq\delta} f_{e\ \gamma} f_{e\ \delta} \sum_{i=1}^{24} Q_{i\ \gamma\delta}\Phi_i \\ &= 8 \sum_{\gamma\neq\delta} f_{e\ \gamma} f_{e\ \delta} (f_{e\ \gamma} f_{e\ \delta} - f^{\perp}_{e\ \gamma} f^{\perp}_{e\ \delta}) = 8 - S_1. \end{aligned} \tag{A.12}$$

Obviously, (A.12) yields (3.12.b).

Appendix B

SOLUTION IN A BOX INCLINED AT $\theta = 45°$

Configurations in Figure 2 with $N_x = n_z$ allow us to simulate a box inclined with an angle $\theta = 45°$ with respect to the x-axis. This is also shown in Figure 3 for a limited box ($N_x = n_z = 1$). Since the x- and z-axes play a symmetric role

$$\Delta_x = -\Delta_z \overset{\text{def.}}{=} \Delta^{(1)}, \ \Delta_{xxz} = -\Delta_{zzx} \overset{\text{def.}}{=} \Delta^{(3)} \tag{B.1}$$

Hence, the time-independent solution (3.11) may be written as

$$\begin{aligned} N_i(\mathbf{r}) &= d(\mathbf{r})e_i^0 + \Delta^{(1)}(\mathbf{r})(C_{i\,x} - C_{i\,z}) + \sum_{\{\alpha,\beta\}\in\{x,z\}} \Delta_{\alpha\beta}(\mathbf{r})Q_{i\,\alpha\beta} + \\ &\quad + \Delta^{(3)}(\mathbf{r})(T_{i\,xxz} - T_{i\,zzx}) + \Delta_{xxzz}(\mathbf{r})T_{i\,xxzz} + \\ &\quad + \Delta^{im}(\mathbf{r})e_i^{im}, \quad i = 1, \ldots, b_m. \end{aligned} \tag{B.2}$$

Then, the invariance conditions (3.14.b) on the populations N_i with $\{i\colon C_{i\,x}C_{i\,z} = 1\}$ or $\{i\colon C_{i\,x} = 0 \text{ and } C_{i\,z} = 0\}$ imply the two following conditions with the help of the solution (3.14.a)

$$\lambda_\psi(\tfrac{1}{2}(\Delta_{xx} + \Delta_{zz}) + \Delta_{xz}) + \frac{\lambda_1}{3}\Delta_{xxzz} = \frac{C^{\text{per.}}}{2}, \tag{B.3.a}$$

$$\lambda_\psi(-\tfrac{1}{2}(\Delta_{xx} + \Delta_{zz})) + \frac{\lambda_1}{3}\Delta_{xxzz} = \frac{C^{\text{per.}}}{2}. \tag{B.3.b}$$

Let us consider first the node z_1 in Figure 3. In this node, one has three bounce-back reflections for populations different from the trivial one (3.8). Bounce-back equations (3.4) hold for directions $k = 1, \ldots, 3$ (see Figure 3)

$$\Delta^{(1)} = \frac{\lambda_\psi}{4}(-1)^k(\Delta_{xx} - \Delta_{zz}) - \frac{\Delta^{(3)}}{2}(2 + \lambda_2) - \frac{\lambda_1}{12}\Delta_{xxzz}$$
$$\text{for } k = 1, 2, \tag{B.4.a}$$

$$\Delta^{(1)} = \frac{3C^{\text{per.}}}{8} + \frac{\lambda_\psi}{4}\left(\frac{\Delta_{zz} + \Delta_{xx}}{2} - \Delta_{xz}\right) + \Delta^{(3)}(2 + \lambda_2) + \frac{\lambda_1}{12}\Delta_{xxzz}$$
$$\text{for } k = 3. \tag{B.4.b}$$

Consequently, from (B.4.a)

$$\Delta_{xx} = \Delta_{zz} \tag{B.5.a}$$

Hence, (B.3.a) and (B.3.b) are subtracted and with the help of (B.5.a)

$$\Delta_{xx} = \Delta_{zz} = -\tfrac{1}{2}\Delta_{xz}. \tag{B.5.b}$$

Hence, the second-order tensors in (B.2) form a perturbation tensor $Q_{i\,nn}(\theta = 45^\circ)$ (cf. (3.6.a) and (3.17.a)). Moreover, one can see that for

$$\lambda_1 = 0, \tag{B.6}$$

Equations (B.5) and (B.3) yield

$$\Delta_{xz} = \frac{C^{\text{per.}}}{\lambda_\psi} \tag{B.7.a}$$

Hence, from (B.4) with (B.5.b), (B.6) and (B.7.a)

$$\Delta^{(3)} = 0, \tag{B.7.b}$$
$$\Delta^{(1)} = 0. \tag{B.7.c}$$

Anomalous currents are eliminated in the last node $\mathbf{z}_l$ by the choice (B.6) of the eigenvalue λ_1 (cf. A.9.b).

The node $\mathbf{z}_2$ in Figure 3 may be considered along the same lines. The only difference with the node $\mathbf{z}_l$ is that two pairs of Boltzmann equations (3.6.b) between the nodes $\mathbf{z}_l$ and $\mathbf{z}_2$ should be considered instead of the bounce-back conditions (3.4) in the x- and z-directions (see populations N_{-1} and N_1'', N_{-2} and N_2'' for node $\mathbf{z}_l$ in Figure 3). Substitution of the solution (B.2) with (B.7) for the node $\mathbf{z}_l$ into these equations leads to the same solution (B.7) in the node $\mathbf{z}_2$ when the eigenvalue (B.6) is used. Subsequently, the solutions (B.6) and (B.7) may be developed at all nodes; hence, the solution (3.15) is obtained for any node in a limited configuration with $\theta = 45°$.

The analysis of the possible solutions in the periodic configuration (Figure 2.d) is similar, but more tedious. Actually, Boltzmann equations (3.6.b) should be solved simultaneously for all the nodes. Such a study for the minimal cell $N_x = n_z = 1$, $N_z = 3$ shows that the common form of the solution in the case $\delta_1 = 0$ is

$$\Delta^{(1)}(\mathbf{z}) = \text{const.}; \tag{B.8.a}$$

$$\Delta^{(3)}(\mathbf{z}) = \text{const.} \qquad \text{for } \lambda_2 \neq -2 \tag{B.8.b}$$

A numerical simulation with $\lambda_2 \neq -2$, $\lambda_1 = 0$ and zero velocity initial distribution converges to the solution (B.7). For $\lambda_2 = -2$, the coefficients $\Delta^{(3)}(z)$, obtained numerically, are not constant in agreement with (B.8.b).

Appendix C

SOLUTION IN A LIMITED BOX INCLINED AT $0° < \theta < 45°$

The purpose of this Appendix is to show that the following solution eliminates the anomalous currents in an arbitrarily oriented limited box

$$\lambda_1 = 0 \tag{C.1}$$

$$\lambda_2 = -2 \tag{C.2}$$

The form of the population solutions is then given by (3.15.c) in any lattice node.

For an arbitrary 2D solution, the populations are given by the relation (3.11) (cf. B.2 with B.1 for $\theta = 45°$). Then, the invariance conditions (3.14.b) with the help of the solution (3.14.a) in the y-direction imply that for the populations (cf. B.3 for $\theta = 45°$)

$$\left(-\frac{\lambda_\psi}{2}(\Delta_{xx} + \Delta_{zz})\right) + \frac{\lambda_1}{3}\Delta_{xxzz} = \frac{C^{\text{per.}}}{2} \tag{C.3}$$

First, the node (1, 1) is considered (see Figure 2.c). At this node, one has four bounce-back reflections of populations different from the trivial one (3.8) in the general case. Then, similar to the relations (B.4), the bounce-back equations (3.4) yield for the directions $k = 1, 2$ (see Figure 3):

$$\Delta_\alpha = (-1)^k \left\{\frac{C^{\text{per.}}}{2} q(\theta) - \frac{\lambda_1}{12}\Delta_{xxzz}\right\} - \frac{\lambda_\psi}{4}(\Delta_{zz} - \Delta_{xx}) - \frac{\Delta_{\alpha\beta\beta}}{2}(2 + \lambda_2) \text{ where } \begin{cases} \alpha = z,\ \beta = x \\ q(\theta) = (\cos^2\theta - \frac{1}{2}) \text{ for } k = 1; \\ \alpha = x,\ \beta = z, \\ q(\theta) = (\sin^2\theta - \frac{1}{2}) \text{ for } k = 2; \end{cases} \tag{C.4.a}$$

Then, the sum and difference of Equations (3.4) for directions $k = 3$ and $k = 4$ (see Figure 3)

$$\begin{aligned} \Delta_x &= \frac{-\lambda_\psi}{2}\Delta_{xz} + \frac{C^{\text{per.}}}{2}\sin 2\theta + \Delta_{xzz}(2 + \lambda_2), \\ \Delta_z &= \frac{-\lambda_\psi}{4}(\Delta_{zz} + \Delta_{xx}) - \frac{\lambda_1}{6}\Delta_{xxzz} - \frac{C^{\text{per.}}}{4} + \Delta_{zxx}(2 + \lambda_2). \end{aligned} \tag{C.4.b}$$

One can see that the solution of Equations (C.3) and (C.4) with the condition (C.1) is

$$\begin{aligned} \Delta_x &= \Delta_z \equiv 0; \qquad \Delta_{xx} = -\frac{C^{\text{per.}}}{\lambda_\psi}\sin^2\theta, \qquad \Delta_{zz} = -\frac{C^{\text{per.}}}{\lambda_\psi}\cos^2\theta, \\ \Delta_{xz} &= \frac{C^{\text{per.}}}{\lambda_\psi}\sin 2\theta. \end{aligned} \tag{C.5}$$

Thus, the form of population (3.15.c) is found by the relations (C.5) for the node (1, 1). Other nodes may be considered along the same lines using the fact that the two Boltzmann equations (3.6.b) between two neighbouring nodes lead to the same solution as with bounce-back in this direction if in one of these nodes the solution (3.15.c) is already verified and the conditions (C.1) are imposed. Then, in order to obtain a solution (3.15.c) in all the nodes (x, z) (Figure 2.c), they may be considered in the following order

$$\begin{aligned}&(1,1),(2,1),(3,1),\dots(N_x-1,1),\\&(1,2),(2,2),\dots(N_x-2,2),\\&\dots\\&(1,n_z),(2,n_z),\dots(N_x-n_z,n_z)\end{aligned}$$

The procedure should be continued backwards using periodic conditions for nodes $(N_x, 1)$ and $(1, n_z)$ (see Figure 2.c):

$$\begin{aligned}&(N_x,1)\\&(1,n_z+1),(2,n_z+1),\dots(N_x-n_z-1,n_z+1)\\&(N_x-1,2),(N_x,2)\\&(1,n_z+2),(2,n_z+2),\dots(N_x-n_z-2,n_z+2)\end{aligned}$$

Consequently, one obtains the solution (3.15.c) in all lattice nodes.

Acknowledgements

A grant from the Institut Français du Pétrole (F. Kalaydjian) which partially supports this work, is gratefully acknowledged.

References

Adler, C., d'Humières, D. and Rothman, D.: 1994, Surface tension and interface fluctuations in immiscible lattice gases, *J. Phys. I, France* **4**, 29–46.

Appert, C. and Zaleski, S.: 1990, A lattice gas with a liquid–gas transition, *Phys. Rev. Lett.* **64**, 1–4.

Appert, C., Rothman, D. H. and Zaleski, S.: 1991, A liquid–gas model on a lattice, *Physica D*, **47**, 85–96.

Appert, C. and Zaleski, S.: 1993, Dynamical liquid–gas phase transition, *J. Phys. II, France* **3**, 309–337.

Appert, C.: 1993, Transition de phase dynamique de type liquide–gaz et création d'interfaces dans un gaz sur réseau, Thèse de Doctorat de l'Université Paris VI.

Burgess, D., Hayot, F. and Saam, W. F.: 1988, Model for surface tension in lattice–gas hydrodynamics, *Phys. Rev. A* **38**, 3589–3592.

Chen, S., Wang, Z., Shan, X. and Doolen, G.: 1992, Lattice-Boltzmann computational fluid dynamics in three dimensions, *J. Stat. Phys.* **68**, 379–400.

Cornubert, R., d'Humières, D. and Livermore, D.: 1991, A Knudsen layer theory for lattice gases, *Physica D*, **47**, 241–259.

Edwards, D. A., Brenner, H. and Wasan, D. T.: 1991, Interfacial transport processes and rheology, *Butterworth–Heinemann Series in Chemical Engineering*.

Frisch, U., Hasslacher, B. and Pomeau, Y.: 1986, Lattice gas automata for the Navier–Stokes equation, *Phys. Rev. Lett.* **56**, 1505–1508.

Frisch, U., d'Humières, D., Hasslacher, B., Lallemand, P., Pomeau, Y. and Rivet, J. P.: 1987, Lattice gas hydrodynamics in two and three dimensions, *Complex Systems*, **1**, 649–707.

Ginzbourg, I. and Adler, P. M.: 1994a, Boundary flow condition analysis for the three-dimensional lattice Boltzmann model, *J. Phys. II, France*, **4**, 191–214.

Ginzbourg, I. and Adler, P. M.: 1994b, Boundary conditions at a plane fluid-interface in the FCHC lattice Boltzmann model, in preparation.

Grunau, D. W.: 1993, Lattice methods for modeling hydrodynamics, PhD Thesis, Colorado State University.

Gunstensen, A. K., Rothman, D. H., Zaleski, S. and Zanetti, G.: 1991, Lattice-Boltzmann model of immiscible fluids, *Phys. Rev. A* **43**, 107–114.

Gunstensen, A. K. and Rothman, D. H.: 1991, A Galilean-invariant two-phase lattice gas, *Physica D* **47**, 53–63.

Gunstensen, A. K. and Rothman, D. H.: 1992, Microscopic modeling of immiscible fluids in three dimensions by a lattice–Boltzmann method, *Europhys. Lett.* **18**, 157–161.

Gunstensen, A.K.: 1992, Lattice-Boltzmann studies of multiphase flow through porous media, PhD Thesis, MIT.

Hayot, F.: 1991, Fingering instability in a lattice gas, *Physica D* **47**, 64–71.

Higuera, F. J. and Jimenez, J.: 1989, Boltzmann approach to lattice gas simulations, *Europhys. Lett.* **9**, 663–668.

Higuera, F. J., Succi, S. and Benzi, R.: 1989, Lattice gas dynamics with enhanced collisions, *Europhys. Lett.* **9**, 345–349.

d'Humières, D., Lallemand, P. and Frisch, U.: 1986, Lattice gas models for 3D hydrodynamics, *Europhys. Lett.* **2**, 291–297.

d'Humières, D. and Lallemand, P.: 1987, Numerical simulations of hydrodynamics with lattice gas automata in two dimensions, *Complex Systems* **1**, 599–632.

McNamara, G. R. and Zanetti, G.: 1988, Use of the Boltzmann equation to simulate lattice-gas automata, *Phys. Rev. Lett.* **61**, 2332–2335.

Rem, P. C. and Somers, J. A.: 1989, Cellular automata on a transputer network, in R. Monaco (ed), *Discrete Kinetic Theory, Lattice Gas Dynamics, and Foundation of Hydrodynamics*, World Scientific, Singapore, pp. 268–275.

Rothman, D. H. and Keller, J. M.: 1988, Immiscible cellular-automaton fluids, *J. Statist. Phys.* **5**, 1119–1127.

Rothman, D. H.: 1990, Macroscopic laws for immiscible two-phase flow in porous media: results from numerical experiments, *J. Geophys. Res.* **B95**, 8663–8674.

Rowlinson J. and Widom, B.: 1982, *Molecular Theory of Capillarity*, Clarendon Press, Oxford.

Somers, J. A. and Rem, P. C.: 1991, Analysis of surface tension in two phase lattice gases, *Physica* **D47**, 39–46.

Succi, S., Foti, E. and Higuera, F.: 1989, Three-dimensional flows in complex geometries with the lattice Boltzmann method, *Europhys. Lett.* **10**, 433–438.

Zanetti, G.: 1991, The hydrodynamics of lattice gas automata, *Physica* **D47**, 30–35.

Transport in Porous Media **20:** 77–103, 1995.

Three-Phase Flow and Gravity Drainage in Porous Media

MARTIN BLUNT, DENGEN ZHOU and DARRYL FENWICK
Department of Petroleum Engineering, Stanford University, Stanford, CA 94305-2220, U.S.A.

(Received: May 1994)

Abstract. We present a theoretical and experimental treatment of three-phase flow in water-wet porous media from the molecular level upwards. Many three-phase systems in polluted soil and oil reservoirs have a positive initial spreading coefficient, which means that oil spontaneously spreads as a layer between water and gas. We compute the thickness and stability of this oil layer and show that appreciable recovery of oil by drainage only occurs when the oil layer occupies crevices or roughness in the pore space. We then analyze the distribution of oil, water and gas in vertical equilibrium for a spreading system, which is governed by $\alpha = \gamma_{ow}(\rho_o - \rho_g)/\gamma_{go}(\rho_w - \rho_o)$, where γ_{ow} and γ_{go} are the oil/water and gas/oil interfacial tensions respectively, and ρ_g, ρ_o and ρ_w are the gas, oil and water densities respectively. If $\alpha > 1$, there is a height above the oil/water contact, beyond which connected oil only exists as a molecular film, with a negligible saturation. This height is independent of the structure of the porous medium. When $\alpha < 1$, large quantities of oil remain in the pore space and gravity drainage is not efficient. If the initial spreading coefficient is negative, oil can be trapped and the recovery is also poor. We performed gravity drainage experiments in sand columns and capillary tubes which confirmed our predictions.

Key words: Three-phase flow, gravity drainage, spreading coefficient.

Nomenclature

c = speed of light (Appendix A)
 = conductance factor (Appendix B)
C_s = spreading coefficient
g = acceleration due to gravity
h = height
H = height of the oil column
J = capillary pressure function
J^* = threshold curvature
P = pressure
P_c = capillary pressure
Q = drainage rate
r = radius of curvature
S = saturation
t = oil-layer thickness
t_d = drainage time
T = temperature
V = oil volume
w = water-layer thickness
z = elevation
z_c = critical height

Greek Letters
α = ratio, Equation (5.11)
γ = interfacial tension
λ = capillary-pressure exponent
Π = disjoining pressure
ρ = density

Subscripts
g = gas
o = oil
w = water

1. Introduction

The displacement of oil by gas in the presence of water is an important recovery process in oil fields and in the cleanup of contaminants spilled below ground. The displacement of oil by gas under gravity (gravity drainage) occurs in oil reservoirs when the gas cap expands as the pressure drops, when oil condensate forms, or when gas is injected into the gas cap. Three-phase flow is also seen when natural gas, nitrogen, carbon dioxide or steam are injected into the field to displace oil. In an environmental context, the spilling and leakage of hydrocarbons and organic solvents are major contributors to groundwater contamination. The low solubility of these products means that they are often present in their own phase. An oil that is less dense than water, such as a fuel, will migrate downwards until it rests above the water table. The effects of water-table movement, capillarity and gravity will smear the oil in a region above and below the water table, where both air and water are also present. Artificial lowering of the water table by pumping results in the displacement of oil by air through a wet soil. This again is a gravity drainage process. As we show in this paper, this can be an effective way to remove non-aqueous phase pollutants.

Since the work of Dumoré and Schols (1974), it has been known that gravity drainage in water-wet rock can lead to a high oil recovery, with residual oil saturations of a few percent in the presence of immiscible gas and water, which is much lower than the residual oil saturation in the presence of water alone. Further studies on sandstone cores, bead packs and sand columns (Chatzis *et al.*, 1988; Kantzas *et al.*, 1988a; Vizika, 1993; Vizika and Lombard, 1994) confirmed these results. It was suspected that the high oil recovery was due to drainage through films of oil that lie between the water and the gas in the pore space. This film drainage has been observed directly in two-dimensional etched glass micromodels by Kantzas *et al.* (1988b), Oren *et al.* (1992), Oren and Pinczewski (1992), Kalaydjian (1992) and Soll *et al.* (1993). It was shown by Oren *et al.* (1992), Vizika (1993) and Kalaydjian (1993) that systems with a positive initial spreading coefficient, which means that the oil spontaneously spreads over a water/gas interface, would experience film drainage and high recoveries, whereas nonspreading systems would see lower recoveries. It was suggested by Kantzas *et al.* (1988b) that the recovery could be determined by the stability of the oil film, which is controlled by capillary and intermolecular forces, rather than the spreading coefficient alone.

This paper investigates the fundamental mechanisms of oil recovery in three-phase flow, in water-wet porous media, starting at the molecular scale, and provides a predictive theory of gravity drainage. The principal issues are: (1) the thickness and stability of thin oil films controlled by intermolecular forces; (2) the thickness of oil layers during drainage; (3) the flow rate in these layers; (4) the final oil recovery and fluid distribution.

We will show that for a spreading system, stable oil layers may form in the pore space, which provide pressure continuity for the oil phase, thus preventing it from being trapped. Oil can drain rapidly by swelling these layers to occupy the crevices and roughness in the pore space. However, the final oil saturation can be essentially zero, with the oil confined to thin, molecular films. In vertical capillary/gravity equilibrium, where the oil density is less than that of water, there is a finite height above the water table (or the water/oil contact) where the connected oil saturation is zero. This height is determined solely by the height of the oil bank, and the interfacial tensions and densities of the fluids, but is independent of the pore size distribution. Nonspreading systems do not allow drainage of oil layers and give lower recoveries from gravity drainage. These findings are confirmed by a series of experiments on sand columns and capillary tubes.

2. Does Oil Spread on Water?

If a drop of oil is placed on a flat liquid substrate its behavior is determined by the initial spreading coefficient C_s:

$$C_s = \gamma_{gw} - \gamma_{go} - \gamma_{ow}, \tag{2.1}$$

γ_{gw}, γ_{go} and γ_{ow} are the gas/water, gas/oil and oil/water interfacial tensions respectively measured on the fluids before they are brought into contact with each other. There are three different things that can happen (Hirasaki, 1993). (i) If $C_s < 0$, the three-phase contact line shown in Figure 1(a) is stable and the drop remains on the liquid surface. Examples include medium- and long-chain alkanes, such as dodecane. (ii) If $C_s > 0$, the contact line between the three phases is unstable and the oil spreads, as shown in Figure 1. This is consistent with everyday experience: gasoline spilled on a puddle of water, for instance, will spread until it forms a thin, iridescent film. Many solvents, hydrocarbons and crude oils (see Table I and p. 104, Table 5 of Muskat (1949)) do have a positive initial-spreading coefficient. We can define an equilibrium spreading coefficient C_s^e from Equation (2.1), where the interfacial tensions are measured in thermodynamic equilibrium. The water surface is coated by a thin oil film, which lowers the effective gas/water surface tension and $C_s^e < 0$. Excess oil remains in a droplet in equilibrium with the film, as shown in Figure 1(b). Many hydrocarbons, such as benzene, and solvents, such

TABLE I. Pure interfacial tensions, initial spreading coefficients and densities for several oil/water/air systems at 20° C. The pure water/air surface tension is 72.9 mNm^{-1}. The water density is 998.2 kgm^{-3} and the air density is 1.2 kgm^{-3}. Data is taken from (Adamson, 1990; Childs, 1949; McBride *et al.*, 1992).

oil	γ_{ow} (mNm^{-1})	γ_{go} (mNm^{-1})	C_s (mNm^{-1})	Density (kgm^{-3})	α Eqn. (5.11)
Benzene	35	28.9	8.7	879	8.8
Gasoline	30	20	3	730	4.1
Hexane	50.8	18.4	3.4	659	5.3
Isoamyl alcohol	5	23.7	44	854	1.2
TCE	35.4	30	7.2	1460	
Carbon tetrachloride	45	27	0.6	1594	

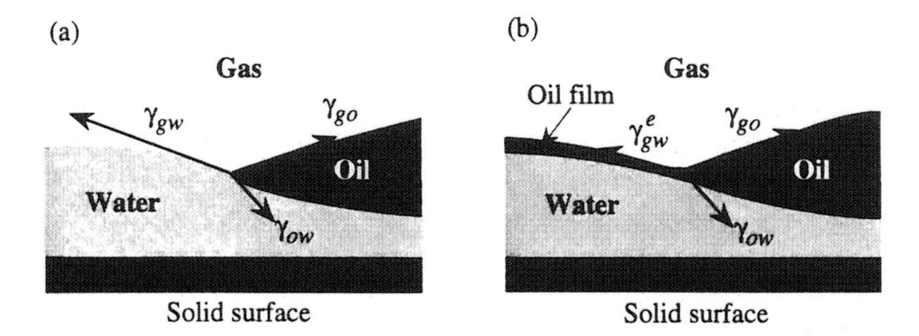

Fig. 1. (a) Three fluids in contact. If $\gamma_{gw} > \gamma_{go} + \gamma_{ow}$ the oil will spontaneously spread between the gas and water. Define a spreading coefficient $C_s = \gamma_{gw} - \gamma_{go} - \gamma_{ow}$. $C_s > 0$ indicates an oil that spreads to form a thin, molecular film on a flat surface. (b) If the oil forms a thin film between the gas and water, the system can be in equilibrium with a macroscopic blob of oil, because of the lowered effective surface tension of the polluted gas/water interface. This film provides pressure continuity of the oil phase, but the drainage rates through it are very small.

as carbon tetrachloride, have a positive initial spreading coefficient, but a negative equilibrium spreading coefficient. The film generally is approximately of molecular thickness, between 0.5 and 5 nm across. (iii) The third possibility is that the oil film can swell without limit. Once the oil film is thicker than the range of intermolecular forces, $C_s^e = 0$ (Adamson, 1990; Gibbs, 1928; Rowlinson and Widom, 1989). An example of this behavior is Soltrol 170, a commercial mixture of hydrocarbons (Oren, *et al.*, 1992; Vizika, 1993).

The thickness of an oil film can be predicted by calculating the van der Waals forces between the water, oil and gas. This computation is described in Appendix A and is similar to the work of Oren and Pinczewski (1991) and Hirasaki (1993), although we use the complete expression for the van der Waals force. We have shown (Blunt *et al.*, 1994) that on a flat surface most hydrocarbons form a film, whose thickness depends on the capillary pressure and displacement history. The

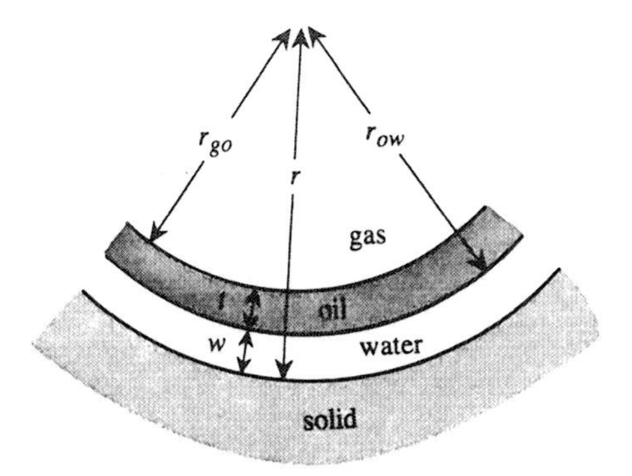

Fig. 2. Three phases in a cylindrical region of the pore space. The oil forms a stable film between the water and gas of thickness t. In our analysis we assume that $r \gg t$. For values of r of a few hundred microns (the upper range of pore sizes in soil and rock), films tens to hundreds of nanometers across can be supported.

same behavior has been predicted for wetting films (Hirasaki, 1988). For capillary pressures that are representative of displacements in porous media, the film thicknesses for both cases (ii) and (iii) are of order a few nanometers or less. The film provides pressure continuity for the oil phase. This means that isolated oil ganglia, trapped and surrounded by water, can become connected during gas injection, when the gas contacts the oil. However, we will show that the drainage rate through these films is far too slow to account for the oil recoveries observed experimentally. Thicker oil layers in crevices of the pore space must provide channels for more rapid drainage.

3. Configuration of Three Phases in the Pore Space

3.1. OIL FILMS IN A CYLINDER

Figure 2 shows the schematic arrangement of fluid in a cylindrical concavity of the pore space. Water wets the solid. Oil is intermediate-wet and occupies a film of thickness t, while the gas, being nonwetting, fills the center of the cylinder. The equilibrium film thickness is found from the augmented Young–Laplace equation:

$$P_{cgo} = \Pi_o(t) + \frac{\gamma_{go}}{r - t - w} = \Pi_0(t) + \frac{\gamma_{go}}{r_{go}} \tag{3.1}$$

where P_{cgo} is the gas/oil capillary pressure, w is the water film thickness, r_{go} is the radius of curvature of the gas/oil interface and $\Pi_o(t)$ is the disjoining pressure that accounts for the influence of intermolecular forces on the oil (Derjaguin and Kussakov, 1939). A positive disjoining pressure is equivalent to a repulsion between the gas and the water, leading to swelling of the oil film, whereas $\Pi_o(t) < 0$ corresponds to an attractive force that makes the film thinner. At distances greater than a few molecular diameters, the main contribution to the disjoining pressure comes from the dispersive, van der Waals force, which gives a small positive $\Pi_o(t)$ for t greater than approximately 10 nm for most alkanes. For t less than around 5 nm, the intermolecular forces are controlled by steric forces, can be very large and are significant compared with the curvature term in Equation (3.1). A similar expression can be written for the pressure difference between the oil and water phases:

$$P_{cow} = \Pi_w(w) - \Pi_o(t) + \frac{\gamma_{ow}}{r_{ow}}, \tag{3.2}$$

where $\Pi_w(w)$ is the disjoining pressure of the water film.

Figure 3 shows $\Pi_o(t) + \gamma_{go}/r_{go}$ plotted as a function of t for an n-octane/water/gas system with a representative water film thickness of 10 nm on a quartz capillary of radius 500 μm. The computation of $\Pi_o(t)$ is described in Appendix A and is accurate if $r_{go} \gg t$. For t less than 10 nm, the intermolecular forces are most significant, whereas when t becomes close to $r - w$, the second term in Eqn. (3.1) diverges. Stable solutions are found when both (3.1) is obeyed and $\mathrm{d}P_{cgo}/\mathrm{d}t < 0$. As indicated in Figure 3, there is a narrow range of capillary pressures, just above γ_{go}/r_{go}, for which a film of thickness between 24 nm and 242 nm is stable, due to the long-range influence of van der Waals forces. At higher capillary pressures, the film collapses to molecular thickness. Such thin films are also observed on convex (protruding) surfaces. If we lower the capillary pressure below γ_{go}/r_{go}, there is no stable solution unless the oil occupies the whole of the cylindrical cross-section and there is no gas present. This spontaneous filling of a pore throat is similar to the snap-off mechanism, which has been described before (Mohanty *et al.*, 1987; Roof, 1970).

The maximum oil film thickness on a cylindrical concavity can be 100s of nanometers. While this is much thicker than a molecular film, it is still more than three orders of magnitude smaller than the radius of curvature of the solid surface.

3.2. OIL LAYERS IN AN ANGULAR CREVICE

In a porous medium, it is also possible for the oil to occupy a thick layer, several microns across in the crevices of the pore space, as shown in Figure 4. This layer is much thicker than the oil films discussed above that are only nanometers across. A thick oil layer can always form if the equilibrium-spreading coefficient is zero or positive. Recently, an energy-balance calculation has shown that oil layers thicker than some critical value are also stable if the equilibrium-spreading coefficient is negative (Dong *et al.*, 1994). The exact condition for stability depends on the layer thickness, the geometry of the pore space and C_s^e. If the condition is not met, the layer collapses to a molecular film if the initial spreading coefficient is positive, and there is no film at all if $C_s < 0$.

In this paper we will consider two extremes: a 'nonspreading' system for which C_s and C_s^e are large and negative and oil films and layers are never observed; and

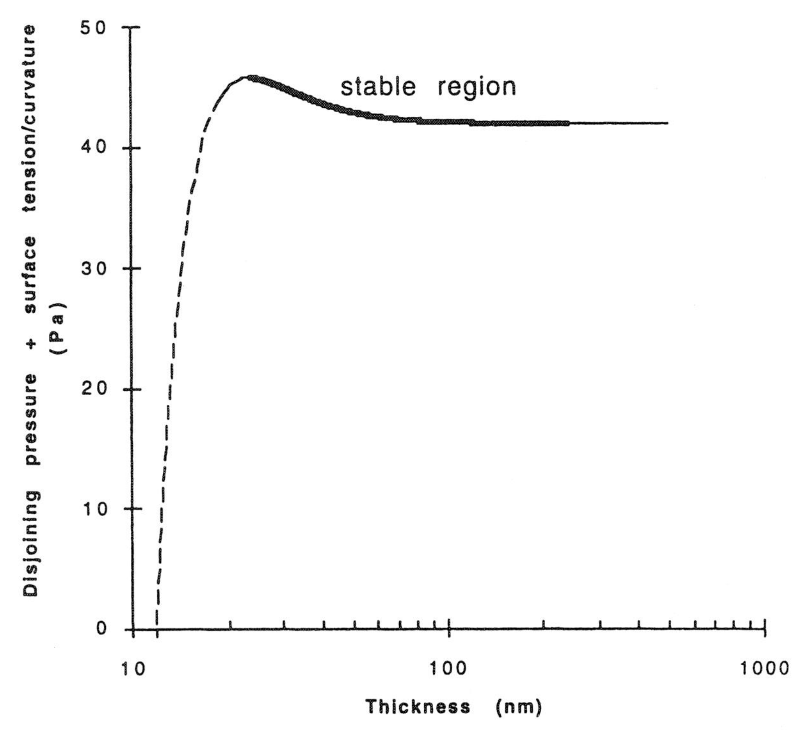

Fig. 3. $\Pi_o(t) + \gamma_{go}/r_{go}$ computed as a function of oil film thickness, t, using the van der Waals intermolecular force calculation described in Appendix A. The fluid system is n-octane, water and air in a capillary of radius 500 μm, which is the same system that will be used in the experiments described later, except that iso-octane will be used rather than n-octane. For a given capillary pressure, P_{cgo}, stable solutions are found when $P_{cgo} = \Pi_o(t) + \gamma_{go}/r_{go}$ and $\mathrm{d}P_{cgo}/\mathrm{d}t < 0$. This region is indicated in bold. A film of thickness between 24 nm and 242 nm can be stable.

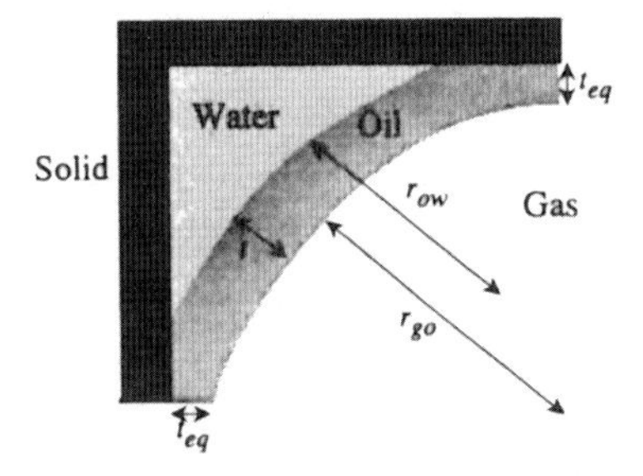

Fig. 4. The arrangement of water, oil and gas in a square crevice. The configuration of fluid shown here is possible if $r_{go} > r_{ow}$, where r_{go} and r_{ow} are the radii of curvature of the gas/oil and oil/water interfaces respectively. In the text we show how to relate this to an inequality on the oil/water and gas/oil capillary pressures. t_{eq} is the thickness of an oil film on a flat surface, typically only a few nanometers across. The layer thickness in a crevice, t, may be much larger, of order the size of the pore space, or several microns.

a 'spreading' system, where C_s^e is positive or very close to zero and oil layers are always stable. In reality there will be a continuous range of behavior dependent on C_s^e.

Figure 4 shows the distributions of oil, water and gas in a square crevice for a spreading system. From geometrical considerations, an oil layer is present if $r_{go} \geq r_{ow}$. From the augmented Young–Laplace Equations (3.1) and (3.2) this means that:

$$\frac{P_{cow} - \Pi_w + \Pi_o}{\gamma_{ow}} \geq \frac{P_{cgo} - \Pi_o}{\gamma_{go}}. \tag{3.3}$$

If we consider thick oil layers, with w and t 100 nm or more, the disjoining pressures will be negligible, in which case the inequality above reduces to:

$$\frac{P_{cow}}{\gamma_{ow}} \geq \frac{P_{cgo}}{\gamma_{go}}. \tag{3.4}$$

If the oil/water pressure difference is much larger than the gas/oil pressure difference, it is possible for a thick oil layer to occupy most of the crevice. A large oil/water capillary pressure forces the water into the corner, while a relatively low gas/oil capillary pressure allows a thick oil layer to develop. This can be true for any angular or sharp groove in the pore space. Since pore sizes are typically in the range of 1 μm to 100 μm or more, oil layers several microns across can exist. If, however, the inequality (3.4), is not obeyed, only a film of molecular thickness will be stable, where the disjoining pressures in Equation (3.3) become significant.

To recap: for a spreading system, on locally flat or convex portions of the pore space, the oil-film thickness is a few nanometers; a cylindrical concavity, with a radius of several hundred microns, can support films tens to hundreds of nanometers across; whereas for sharp or angular crevices, layers several microns thick will form if the oil/water capillary pressure is much larger than the gas/oil capillary pressure.

4. Drainage Rates

4.1. PREDICTED RATES

Figure 5 shows the vertical arrangement of water, oil and air (gas) in a capillary tube. There are two oil ganglia separated by an air bubble of height h. If there is a film of oil that connects the two blobs, there will be pressure continuity in the oil phase which allows the upper ganglion to drain into the lower one under gravity. Since the air has a low density, the pressures in the two ganglia are approximately equal when drainage starts. The gas/oil interfaces at $z = 0$ and $z = h$ are hemispherical caps of radius r, with a total curvature of $2/r$. Thus the gas/oil capillary pressure is approximately $2\gamma_{go}/r$.

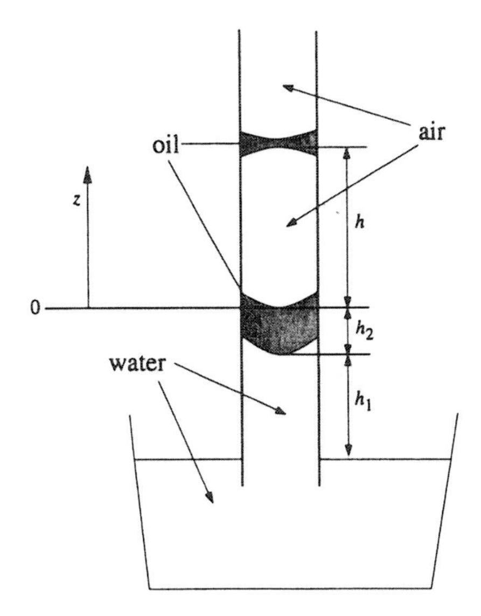

Fig. 5. Fluid configuration in a capillary tube at the beginning of drainage. Two oil ganglia are separated by a gas bubble of height h. The upper ganglion can drain if there is pressure continuity in the oil phase, which means that there must, at least, be an oil film present initially.

In a capillary tube of square cross-section, the fluid configuration is shown in Figure 4. If $P_{cgo} = 2\gamma_{go}/r$, then the gas/oil radius of curvature, $r_{go} = r/2$. Water will also occupy the corner, but as we increase z, the oil/water capillary pressure will rise, forcing water further into the corner. Thus, for large values of h, the oil can occupy almost all the corner of the tube, resulting in rapid drainage. For smaller values of h, the water occupies more of the corners and the drainage rate is lower. In Appendix B we perform a calculation to find the drainage rates in capillaries of square and circular cross-section.

Despite its simple geometry, the analysis for a cylindrical capillary is less straightforward. There is no obvious force balance on the oil film and the oil pressure is not easily determined. $P_{cgo} = 2\gamma_{go}/r$ implies a thin, molecular film and a negligible drainage rate. However, when the tube is held upright, gravity forces instantaneously lead to an increase in oil pressure. Moreover, small vibrations can perturb the pressure in the oil and gas phases. γ_{go}/r is only 42 Pa in the experiments we perform, compared with the atmospheric pressure of 10^5 Pa. To match the experimental results below, it appears that the oil pressure rises to make $P_{cgo} \approx \gamma_{go}/r$ and the oil film swells to reach its maximum stable thickness.

4.2. EXPERIMENTAL CONFIRMATION

We tested our predictions for the drainage times by performing a series of experiments with glass capillary tubes. We first filled the tube with water. Then oil was introduced into the tube at one end to form the lower oil blob. For runs without a

TABLE II. Fluid properties for the experiments presented here. The non-spreading oil is Drakeol 5. For the spreading systems, oil and 'water' phases were made from equilibrium mixtures of iso-octane, distilled water and iso-propanol, where the oil/air surface tension is 21 mNm^{-1}. The data for the oil/water interfacial tension is taken from (Morrow *et al.*, 1988).

Expt.	γ_{ow} (mNm^{-1})	$(\rho_w - \rho_o)$ (kgm^{-3})	ρ_o (kgm^{-3})	μ_o (kgm^{-1}s^{-1})	α Eqn. (5.11)
Non-spreading mineral oil, Expt. 6	54	157	833	0.005	8.9
Fig. 6, Expt. 1	38.1	310	690	0.00048	4.03
Expt. 2	18.47	310	690		1.95
Expt. 3	11.64	310	690		1.23
Fig. 7, Expt. 4	4.41	305	690	0.00049	0.47
Expt. 5	2.03	300	695		0.22

lower oil ganglion ($h_2 = 0$) this step was omitted. Water was subsequently drained out of the opposite end to allow air to enter the tube. By controlling the amount of water that drained out, we controlled the height of the gas bubble, h. The last step was to allow some oil into the tube, above the gas. The tube was then placed vertically and we recorded the time for all the oil in the upper ganglion to drain.

We conducted experiments for high (iso-octane and distilled water) and low (a mixture of iso-octane, water and iso-propanol partitioned into two phases) oil/water interfacial tensions. The fluid properties are shown in Table II. The fluids partially oxidize in air, altering their interfacial tensions from their pure values. The experiments were performed on fluids left exposed to the air for two hours or more. We measured the interfacial tensions periodically for a week and noticed no change after two hours. From our measurements of interfacial tension on oxidized fluids (shown in Table II), both systems had positive initial spreading coefficients. The equilibrium spreading coefficients for the fluids are not known. From the results below it is clear that stable oil layers are formed, which implies that C_s^e is close to zero, and we have a spreading system.

For a capillary tube of circular cross-section with $r = 500\,\mu$m, $h_2 = 0$ and $h = 2$ cm, it took three weeks to drain 0.8 mm^3 of iso-octane from the upper ganglion. The uncertainty in our measurement of the oil volume was ± 0.4 mm^3. This is the same system for which we performed the intermolecular force calculation in the previous section. An oil film of thickness $t = 310 \pm 70$ nm would give this drainage rate, using Equation (B.2) in Appendix B. This will overestimate t, since we have ignored any lubrication effect due to the flow of the water film. Our predicted oil-film thickness from the previous section is 242 nm, which is consistent with our measurements.

We then performed a series of experiments in a capillary tube of square cross-section with a side of length 150 μm. We repeated the experiment for various

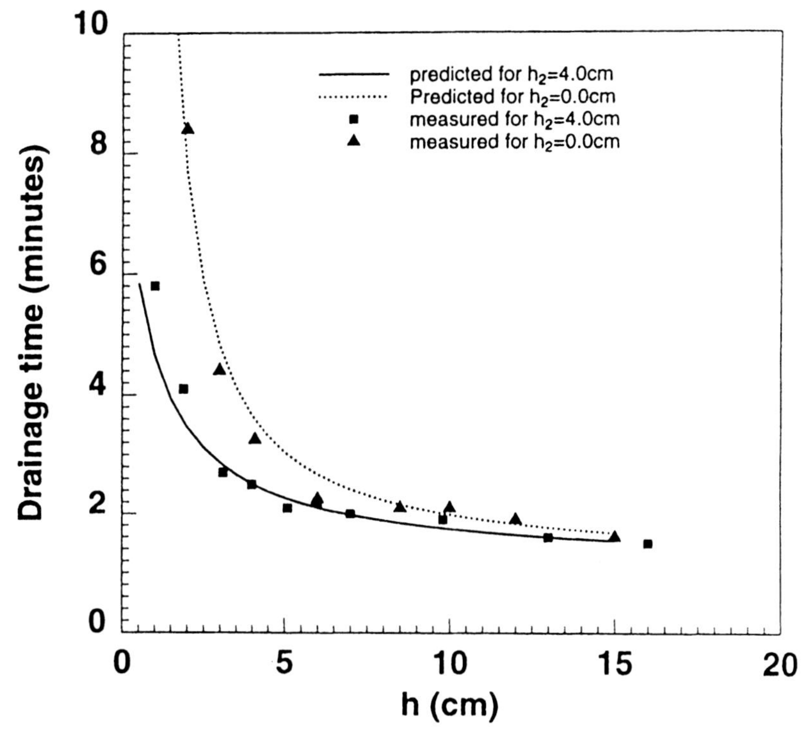

Fig. 6. Comparison of measured and predicted drainage times in a capillary tube of square cross-secton for a system with a high oil/ water interfacial tension.

values of h_2 and h. In each experiment the volume of oil V above the gas was $0.225\,\mathrm{mm}^3$.

Figure 6 shows the results for the system with a high interfacial tension (IFT). The triangles are the drainage times for different values of h when there is no initial oil bank ($h_2 = 0$) and the squares are the drainage times for the same fluids for various h and $h_2 = 4\,\mathrm{cm}$. Increasing h reduces the drainage time. The solid and dotted curves are our predicted times. The only unknown parameter in the equations is c, the conductance constant for the oil layer (see Appendix B). This was estimated to match the experimental results – only one parameter was used to match both curves. The agreement between experiment and theory is good. Notice that the minimum drainage time is just a few minutes, or several thousand times faster than in the cylindrical capillary tube, even though the tube is smaller for this experiment. The reason for this is that the oil can form a much thicker layer (up to $15\,\mu\mathrm{m}$ across) in the corners of the tube than can be supported on a smooth, concave interface (only around 200–300 nm).

Figure 7 shows the results from the low IFT system in the same capillary tube. Again the agreement between experiment and theory is good. The drainage times

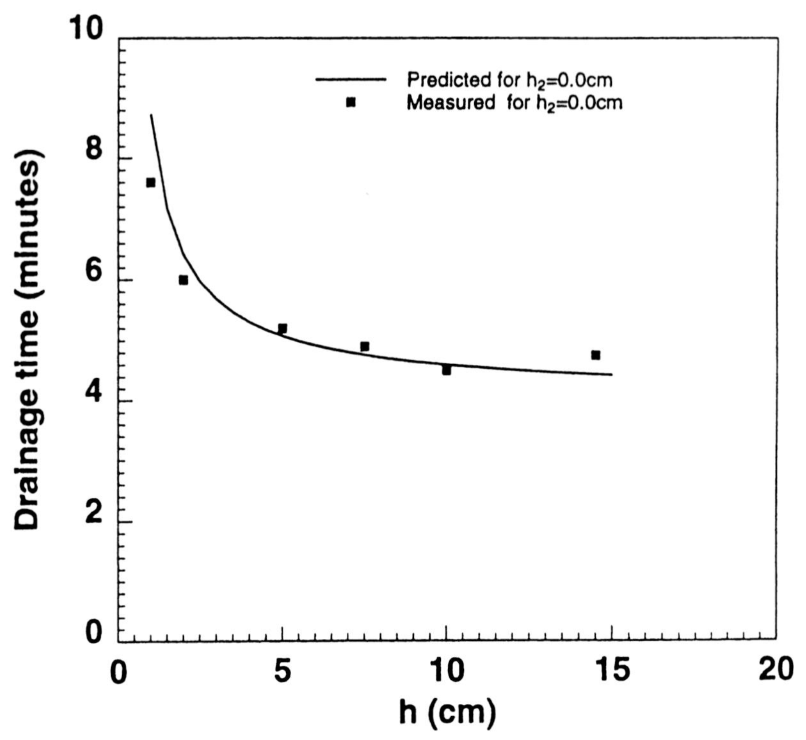

Fig. 7. Comparison of measured and predicted drainage times in a capillary tube of square cross-section for a system with a low oil/water interfacial tension. The drainage times are longer than shown in Figure 6, since the aqueous phase viscosity is $2.8 \times 10^{-3}\,\mathrm{kgm^{-1}s^{-1}}$, rather than $10^{-3}\,\mathrm{kgm^{-1}s^{-1}}$ for the high IFT system.

are longer (and the value of c is smaller) because the viscosities of the water and oil phases are higher than for the high IFT fluids.

We performed the same experiments with a mineral oil that has a negative initial spreading coefficient. Again the reported interfacial tension values were measured once the oil had been exposed to the air for several hours. In this case there was no drainage of the oil, since a spreading film was never established. On square capillaries with a side of 1 mm or larger, droplets of the oil were observed to fall down the glass, like rain droplets on a window pane. However, this phenomenon is only seen for large droplets and large capillaries and is unlikely to be significant in porous media.

Real porous media do contain angular or sharp crevices that can support thick oil layers of order microns across during gravity drainage. This provides a mechanism for relatively rapid displacement of oil. Flow rates across flat surfaces or on uniformly concave interfaces are much slower. In all cases there is agreement between the predicted and measured flow rates.

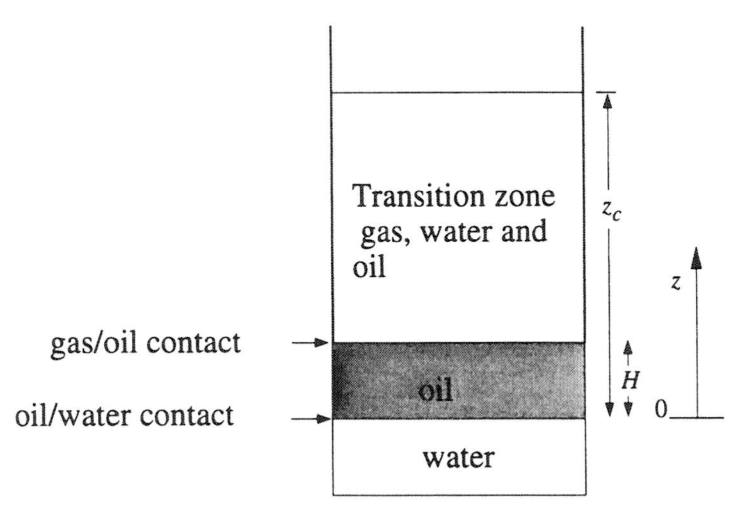

Fig. 8. A schematic of the arrangement of water, oil and air in vertical equilibrium. We define $z = 0$ as the level when oil is first connected. $z = H$ represents the height when gas is first connected. Above this all three phases may be continuous. For most fluid systems there is a critical height, z_c, beyond which connected oil can only be present as a thin film, which means that the oil saturation is essentially zero. This height depends only on H and the fluid properties and is independent of the soil or rock type.

5. Vertical Equilibrium

5.1. A CRITICAL HEIGHT

We will now analyze the fluid distribution in vertical capillary/gravity equilibrium, at the end of drainage for oils less dense than water. Consider again the arrangement of fluid illustrated in Figure 4. If we assume that we have a spreading system where the oil layers are always stable, there is no direct contact of water by gas. To the gas phase, oil and water combined appear to be the wetting phase. This means that the gas/oil capillary pressure can be represented as a function of total liquid saturation $(S_o + S_w)$, as first suggested by Leverett (1941) and confirmed by Parker *et al.* (1987). In contrast, when $C_s < 0$, oil remains in the system as lenses and is influenced by both oil and water separately. As shown by Kalaydjian (1992), this results in a gas/oil capillary pressure that is a function of both S_w and S_o rather than $S_o + S_w$ alone.

Figure 8 shows gas, oil and water in vertical equilibrium. $z = 0$ is defined as the level at which oil is first mobile, or continuous, through the soil or rock. $z = H$ corresponds to the height at which gas is first continuous. Above $z = H$ all three phases may be continuous. This diagram corresponds to the arrangement of non-aqueous phase pollutant resting on the water table, or oil and gas in a reservoir. Where the phases are connected we can write down expressions for the pressures as a function of height:

$$P_w = -z\rho_w g, \tag{5.1}$$

where P_w is the water pressure, g the acceleration due to gravity and we define $P_w = 0$ at $z = 0$. Similarly we may write:

$$P_o = P^*_{cow} - z\rho_o g, \tag{5.2}$$

$$P_g = P^*_{cgo} + P^*_{cow} - H\rho_o g - (z - H)\rho_g g, \tag{5.3}$$

where P^*_{cow} and P^*_{cgo} are the threshold capillary pressures for oil invasion into water and gas invasion into oil respectively.

The capillary pressures are:

$$P_{cow} = P_o - P_w = P^*_{cow} + z(\rho_w - \rho_o)g, \tag{5.4}$$

$$P_{cgo} = P_g - P_o = P^*_{cgo} + (z - H)(\rho_o - \rho_g)g. \tag{5.5}$$

If oil layers prevent any direct gas/water contact, the gas/oil capillary pressure is a function of the sum of the water and oil saturations. The oil/water capillary pressure, where water is the wetting phase, is a function of S_w. We assume that the functional forms of both capillary pressures are the same, but multiplied by their respective interfacial tensions, which control the relative strength of capillary forces (Bear, 1972; Leverett, 1941).

$$P_{cow} = \gamma_{ow} J(S_w) + \Pi_w - \Pi_o, \tag{5.6}$$

$$P_{cgo} = \gamma_{go} J(S_w + S_o) + \Pi_o, \tag{5.7}$$

where J is a capillary pressure function that represents the curvature of the fluid interfaces as a function of saturation. In terms of the microscopic configuration of fluid, shown in Figure 4, $J(S_w)$ in Equation (5.6) is $1/r_{ow}$ and $J(S_w + S_o)$ is $1/r_{go}$. However, this argument is completely general and does not rely on any particular model of the pore-level fluid distribution. The threshold capillary pressures can be written $P^*_{cow} = \gamma_{ow} J^*$ and $P^*_{cgo} = \gamma_{go} J^*$, where J^* is the curvature necessary for a phase to first enter the porous medium.

The capillary pressure decreases with wetting phase saturation (Bear, 1972). Hence:

$$J(S_w) \geq J(S_w + S_o). \tag{5.8}$$

This is equivalent to stating that $r_{go} \geq r_{ow}$ for an oil layer to exist in Figure 4. From Equations (5.6) and (5.7) the inequality above becomes:

$$\frac{P_{cow} - \Pi_w + \Pi_o}{\gamma_{ow}} \geq \frac{P_{cgo} - \Pi_o}{\gamma_{go}}. \tag{5.9}$$

Notice that this is identical to Equation (3.3). By using a capillary pressure analysis, or by considering the microscopic arrangement of fluid in a pore, we arrive at the same inequality for continuity of the oil phase. We substitute Equations (5.4) and (5.5) into (5.9) to find:

$$\frac{z - \frac{\Pi_w - \Pi_o}{(\rho_o - \rho_g)g}}{(z - H) - \frac{\Pi_o}{(\rho_w - \rho_o)g}} \geq \alpha, \tag{5.10}$$

where

$$\alpha = \frac{\gamma_{ow}(\rho_o - \rho_g)}{\gamma_{go}(\rho_w - \rho_o)}, \tag{5.11}$$

α is a property of the interfacial tensions and densities. This expression, without accounting for disjoining pressure, was first derived by Kantzas *et al.* (1988b).

For a bulk phase to be present, the disjoining pressures will be negligible and we may write:

$$\frac{z}{z - H} \geq \alpha. \tag{5.12}$$

For $0 < \alpha < 1$, the inequality above is always obeyed, which means that connected oil exists at all heights above the oil bank. If $\alpha > 1$, there is a finite height at which oil in thick layers cannot exist, which means that oil must reside in thin films a few nanometers across, where the disjoining pressures are significant. The oil saturation of this film will be at most 0.01% and may be considered negligible. The critical height z_c at which the oil saturation becomes virtually zero is:

$$z_c = \frac{\alpha H}{\alpha - 1} = H + \frac{H}{\alpha - 1}. \tag{5.13}$$

For systems with $\alpha > 1$ the minimum oil saturation is zero. In contrast, residual oil saturations in the range 0.1 to 0.5 are encountered in water-saturated porous media (Dullien, 1992). Lowering the water table in a region polluted by free product, or gas cap expansion into a water-flooded reservoir, will mobilize this trapped oil and allow some of it to be recovered by direct pumping.

Figure 9 shows schematic graphs of saturation versus height for different values of α. Figure 9(c), for $\alpha > 1$, demonstrates how the connected oil saturation decreases to zero at some critical height z_c. Above z_c, the fluid distribution is governed by the gas/water capillary pressure (the sum of Equations (5.4) and (5.5)). The gas/water interfaces will be polluted with an oil film, giving a lowered effective gas/water surface tension and an effective spreading coefficient that is approximately zero for many systems (Adamson, 1990; Gibbs, 1928; Rowlinson and Widom, 1989). If the saturation is continuous at $z = z_c$ the effective gas/water surface tension must be $\gamma_{ow} + \gamma_{go}$, which corresponds to $C_s^e = 0$.

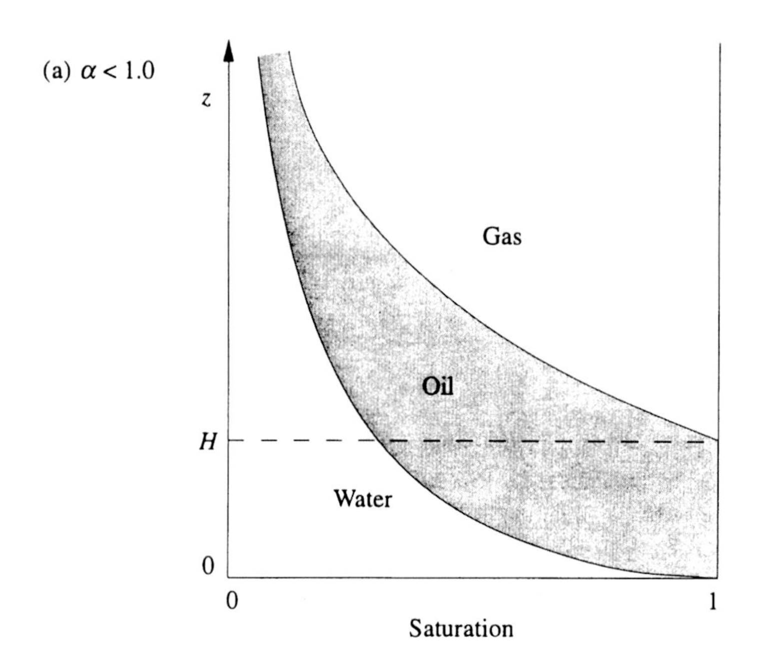
(a) $\alpha < 1.0$
z
Gas
Oil
H
Water
0
0
1
Saturation

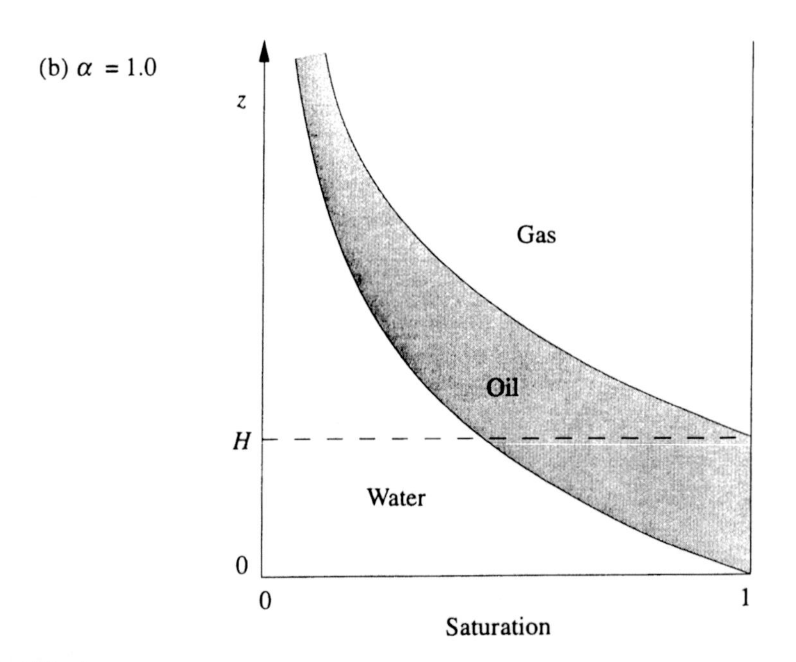
(b) $\alpha = 1.0$
z
Gas
Oil
H
Water
0
0
1
Saturation

Fig. 9.

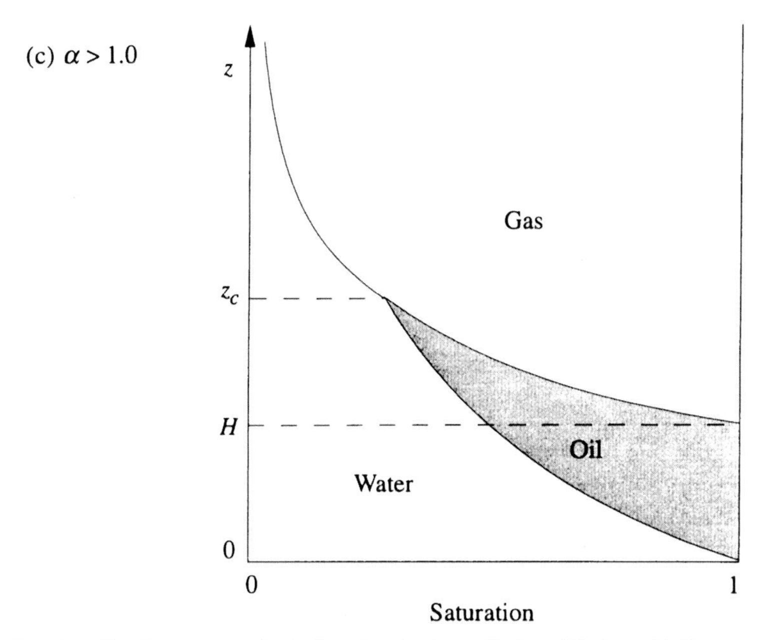

Fig. 9. The distribution of oil, air and water in vertical equilibrium. (a) For $\alpha > 1$, oil remains in the pore space for all z. (b) For $\alpha = 1$, oil just persists at all heights. (c) For $\alpha < 1$, there is a finite height z_c at which there is no connected oil.

5.2. HOW MUCH OIL CAN BE TRAPPED?

We have asserted that the minimum oil saturation can be zero. This analysis is correct if the oil always forms a stable layer between gas and water. This is only true for $C_s^e = 0$. If the equilibrium-spreading coefficient is negative, there is a minimum stable oil-layer thickness (Dong *et al.*, 1994). At low saturations, the oil layers will spontaneously retract, if $C_s^e < 0$, allowing oil to be trapped. Moreover, some oil may sorb to the solid surfaces or be retained in ganglia connected by thin layers or films which never reach gravity equilibrium, because of the very slow drainage rates through them. Hence, we might expect a continuous change in residual oil saturation with C_s^e, from almost zero for $C_s^e \geq 0$ to values close to the residual saturation in water if C_s^e is large and negative.

If gas displaces connected oil, there is no mechanism for the oil to become trapped, apart from the retraction of thin layers discussed in the previous paragraph, and the minimum oil saturation for a spreading system is very low. If gas displaces trapped oil, such as water-flood residual oil in a reservoir, or immobile product just below the water table, oil can remain trapped if it has not been directly contacted by gas. This interpretation is consistent with sand column experiments which showed better recoveries for gravity drainage from continuous oil than for drainage of hitherto residual oil (Kantzas *et al.*, 1988a).

In this section we will consider gravity drainage of previously discontinuous oil for systems with $C_s^e \geq 0$. An indication of the amount of oil that can remain trapped is the water saturation at z_c. Trapped oil at $z = z_c$ is contained in ganglia completely surrounded by water that has not been displaced by gas. At heights above and below z_c less oil will be trapped. Above z_c, the water saturation is lower and more oil will have been contacted by gas. Below z_c the mobile oil will have reconnected previously trapped ganglia.

If we know the three-phase capillary pressures, we can calculate the oil and water saturations. One possible parameterization for the capillary pressure function J is (Brooks and Corey, 1966):

$$J(S) = J^* S^{-1/\lambda}, \tag{5.14}$$

where J^* represents the threshold entry curvature (Corey, 1986), S is the wetting phase saturation and λ is a constant that depends on the pore structure of the medium and is generally in the range 0.2–1.0 (Lenhard and Parker, 1990). Other expressions for the capillary pressure have been proposed (Genuchten, 1980; Parker *et al.*, 1987). The capillary pressure we use has no irreducible or residual water saturation (Lenhard and Parker, 1990). We could allow an irreducible water saturation, but we do assume that all the oil-filled pores can be accessed by gas.

We use Equations (5.4), (5.6), (5.11) and (5.14) to find the water saturation as a function of height:

$$S_w = \left(1 + \frac{z(\rho_o - \rho_g)g}{\alpha J^* \gamma_{go}}\right)^{-\lambda} \tag{5.15}$$

and at $z = z_c$, from Equation (5.13):

$$S_w(z_c) = \left(1 + \frac{H(\rho_o - \rho_g)g}{(\alpha - 1)J^* \gamma_{go}}\right)^{-\lambda} \tag{5.16}$$

If we increase H, the water saturation at z_c decreases, as illustrated in Figure 10. This means that the trapped oil saturation at z_c decreases and the gravity drainage process is more efficient. If we take typical values for a polluted sandy soil: $H = 0.1$ m, $J^* = 10^4\,\mathrm{m}^{-1}$, $\alpha = 4$, $\lambda = 1$, $\rho_w - \rho_g = 10^3\,\mathrm{kgm}^{-3}$, $\rho_o - \rho_g = 700\,\mathrm{kgm}^{-3}$ and $\gamma_{go} = 0.02\,\mathrm{Nm}^{-1}$, we find $S_w = 0.47$, which could allow some oil to be trapped. However, if the pore size distribution is uniform, very little oil remains trapped, even for small values of H, as demonstrated in sand column experiments (Kantzas *et al.*, 1988a). In contrast, a large oil bank in a water-wet reservoir, with the same values as above, except $J^* = 10^5\,\mathrm{m}^{-1}$ and $H = 100$ m, gives S_w as less than 1%. The residual oil saturation at z_c must therefore be much less than 1%. Oil is only trapped if we have a mixed-wet or oil-wet system.

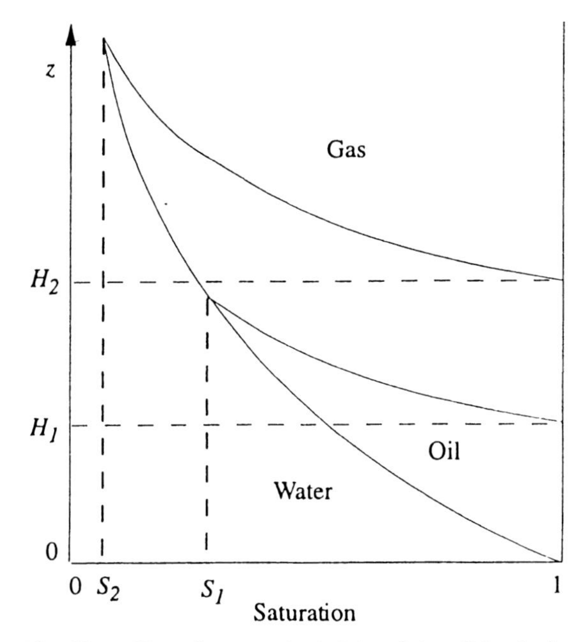

Fig. 10. If we increase the height of the oil bank, from H_1 to H_2, the water saturation at which connected oil no longer exists decreases from S_1 to S_2, as shown. If gas displaces discontinuous oil, oil will remain trapped if it is not contacted directly by gas, because of water blocking. As we increase H, the water saturation at the critical height decreases and water blocking becomes less significant.

5.3. EXPERIMENTAL CONFIRMATION

Several investigators have performed gravity drainage experiments, where gas displaces water and residual oil, or oil and residual water, under gravity (Blunt *et al.*, 1994; Chatzis *et al.*, 1988; Dumoré and Schols, 1974; Kantzas *et al.*, 1988a, b; Vizika, 1993; Vizika and Lombard, 1994) and have shown that final oil saturations as low as 1% are possible. These very good recoveries have been explained by the drainage of oil layers between the gas and water (Kalaydjian, 1992; Kalaydjian *et al.*, 1993; Oren *et al.*, 1992; Oren and Pinczewski, 1992; Soll *et al.*, 1993). It has been shown that lower final oil saturations are seen for systems with a positive spreading coefficient than for nonspreading oils (Kalaydjian *et al.*, 1993; Oren *et al.*, 1992; Vizika, 1993).

We performed experiments in sand columns of two heights: a long column of 97.5 cm, and a short column of 47 cm. Both columns had a diameter of 2 cm. They were filled with a clean, well-sorted sand with a mean grain diameter of approximately 0.3 mm, a permeability of 48 D and porosity of 0.28.

The sand column was first fully saturated with distilled water. The top and bottom valves at the end of the column were then both opened to allow the invasion of air and the free drainage of water. We waited at least 24 hours until no further water was produced. 30 cm^3 of oil was then slowly poured into the top of the column

TABLE III. Summary of experimental measurements

Expt.	1	2	3	4	5	6 (non-spreading)
α (Equation (5.11))	4.03	1.95	1.23	0.47	0.22	8.9
NAPL in long column (cm^3)	6.5	9.0	10.0	14.0	12.0	14.0
NAPL in short column (cm^3)	6.0	8.0	8.5	8.0	6.0	9.0
Difference (cm^3)	0.5	1.0	1.5	6.0	6.0	5.0

to represent the migration of pollutant towards the water table. Oil accumulated at the bottom of the column and was allowed to drain out freely. Periodically, air at just above atmospheric pressure was pumped into the top of the column to displace the oil bank. This exercise stopped when there was no further production of fluids. No further oil was recovered after two weeks of drainage.

For a mineral oil with a negative initial spreading coefficient, (Drakeol 5), we found that 14 cm^3 remained in the long column and 9 cm^3 remained in the short column. We also conducted the experiment described here for systems with a positive initial spreading coefficient and various values of α. By using Corey-type capillary pressures, (5.14), the amount of oil left in the columns was predicted using $\lambda = 0.92$, a value previously measured on a well-sorted sand by Lenhard and Parker (1990) was found by assuming the air/water capillary pressure was zero at the base of the column. J^* was calculated by measuring the capillary rise of water in the column after drainage in air (which was 20.5 cm).

Table III shows the results of the experiments. There is little change in oil saturation with α for the short column. This is because the critical height z_c is above the top of the column for most of the experiments, and thus it is difficult to distinguish between $\alpha < 1$, $\alpha > 1$, spreading and non-spreading systems. There is a systematic change in oil recovered with α for the long column. The difference in recovery for the two columns represents the average saturation in the upper portion of the long column which varies from $14 \pm 1\%$ for $\alpha = 0.47$ to $1 \pm 1\%$ for $\alpha = 4.0$. This difference is plotted in Figure 11. For a spreading system, the average oil saturation above the oil/water contact changes by more than an order of magnitude with α. The trend with α is predicted successfully, as shown in Figure 11. The predictions and measurements are not expected to match exactly, since we used a very crude representation for the capillary pressure. For large α the amount left is zero to within experimental error. The non-spreading oil (the point on the right in Figure 11) gives a poor recovery, even though it has a large value of α, because in this case oil can be trapped.

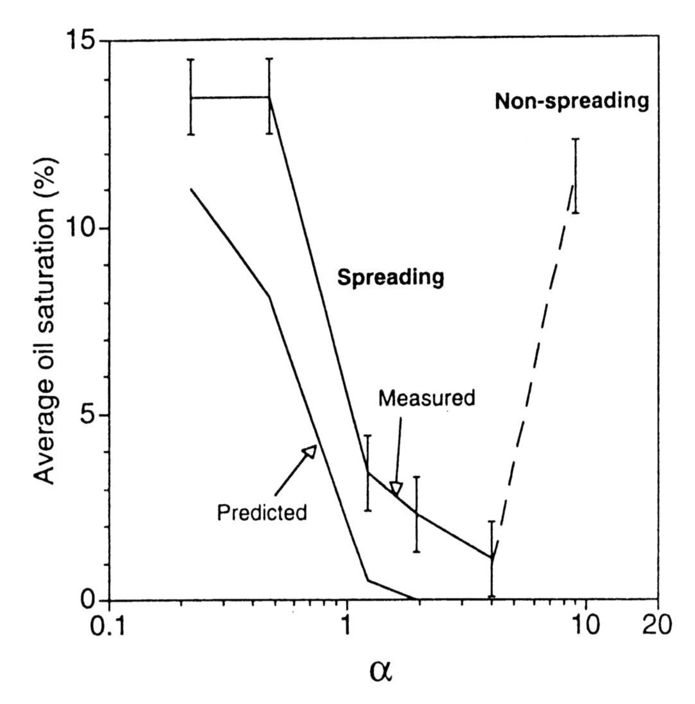

Fig. 11. The average oil saturation in the upper half of the long sand column as a function of α. The experimental error in the measurements is $\pm 1\%$. For spreading systems the saturation can be predicted from first principles using Corey-type capillary pressures. The points connected by solid lines are for spreading systems. The point on the right is for a nonspreading mineral oil. For spreading systems there is a dramatic drop in the amount of oil remaining when α increases beyond 1. For a nonspreading oil, the recovery is poor even for large α.

We measured the oil saturation at the top of the column directly. After three weeks of drainage in the iso-octane system (with $\alpha = 4$) we took samples of the sand and resident fluids from the long column. The samples were washed with iso-propanol to dissolve both the remaining octane and water. The fluid was then run through a gas chromatograph. Above the critical height the oil saturation varied from 1.0 to $3.9 \pm 0.1\%$. This is in the range of the minimum oil saturations between 1 and 3% found previously from gravity drainage experiments (Kantzas *et al.*, 1988a, b) and confirms that low oil saturations are possible under favorable conditions. However, the saturation is not zero and there is some trapping of oil. This could be because the equilibrium spreading coefficient is slightly negative.

5.4. DISCUSSION

The critical parameter that determines oil recovery by gravity drainage for a spreading system is α, which is a property of the fluid system alone and independent of the porous medium, as long as it is water-wet. For most fluids $\alpha > 1$, which means that above a critical height the residual oil saturation can be zero. However, if surfactant flooding is used to displace oil in the presence of gas, this will dramatically decrease γ_{ow}, and α will be less than 1. This will mean that in vertical equilibrium, appreciable quantities of oil can be retained above the oil/water contact. The recovery of oil from gravity drainage is most efficient for large α, which can be achieved by lowering the gas/oil surface tension. In oil reservoirs, the natural gas may be almost miscible with the oil, particularly in condensate reservoirs, leading to very low values of $\gamma_{go}/(\rho_o - \rho_g)$ and an extremely high oil recovery.

6. Conclusions

The mechanism of gravity drainage is transport through thick oil layers sandwiched between water and gas, which occupy the crevices of the pore space. In our experiments these layers allowed rapid drainage for spreading systems, whereas the fluids with a negative initial spreading coefficient did not form oil layers and gave poorer recoveries.

The distribution of oil, water and gas in vertical equilibrium for a spreading system is controlled by the parameter α, Equation (5.11). Typically $\alpha > 1$, and there is a finite height above which the oil saturation can be almost zero. This height is a function only of the fluid densities and interfacial tensions and is independent of the soil or rock type. For $\alpha < 1$, which is seen for surfactant floods, a large quantity of oil can be contained above the oil bank.

Appendix A. Computation of Disjoining Pressure

Using quantum-field theory, and assuming that the film thickness, t, is much greater than a molecular diameter, the following expression for the disjoining pressure has been derived (Dzyaloshinskii *et al.*, 1961; Lifshitz, 1955; Mahanty and Ninham, 1976):

$$\Pi_o(t) = -\frac{kT}{\pi c^3} \sum_{n=0}^{\infty'} \xi_n^3 \varepsilon_2^{3/2} \int_1^{\infty} (I_1 + I_2) p^2 \, \mathrm{d}p, \tag{A.1}$$

where

$$\begin{aligned} I_1 &= \left(\frac{1}{\Delta_{21}\Delta_{32}} \exp\left(\frac{2p\xi_n}{c} t\sqrt{\varepsilon_2} \right) - 1 \right)^{-1}, \\ I_2 &= \left(\frac{1}{\bar{\Delta}_{21}\bar{\Delta}_{32}} \exp\left(\frac{2p\xi_n}{c} t\sqrt{\varepsilon_2} \right) - 1 \right)^{-1}, \\ \Delta_{21} &= \frac{\Delta_{24} + \Delta_{41}\exp(-ws_4/t)}{1 + \Delta_{24}\Delta_{41}\exp(-ws_4/t)}, \quad \Delta_{ij} = \frac{\varepsilon_i s_j - \varepsilon_j s_i}{\varepsilon_i s_j + \varepsilon_j s_i}, \\ s_k &= \sqrt{\varepsilon_k/\varepsilon_2 - 1 + p^2}, \quad \bar{\Delta}_{ij} = \frac{s_j - s_i}{s_j + s_i}, \end{aligned} \tag{A.2}$$

and we consider an oil film (medium 2) between gas (medium 3) and a water film (medium 4) of thickness w absorbed onto a solid surface (medium 1). $\varepsilon_j (j = 1, 4)$ are the dielectric responses as functions of the imaginary frequency $i\xi_n$, where $\xi_n = 2\pi kT/\hbar$ $(n = 1, 2, 3, \ldots)$, k is Boltzmann's constant and $2\pi\hbar$ is Planck's constant. c is the speed of light in a vacuum and T is the absolute temperature, which we set to 300 K in our calculations. The summation is over integral values of n, and the prime on the summation means that the first term in the sum should be multiplied by 1/2. The functions $\varepsilon_1, \varepsilon_2$ and ε_4 are computed from the adsorption spectra of quartz, oil and water, respectively. $\varepsilon = 1$ for gas.

The dielectric response functions are calculated using the Lorentz harmonic oscillator equation:

$$\varepsilon(i\xi_n) = 1 + \sum_r \frac{C_r}{1 + \frac{\xi_n}{\omega_r}} + \sum_j \frac{f_j}{1 + \frac{\xi_n^2}{\omega_j^2} + g_j \frac{\xi_n}{\omega_j^2}}. \tag{A.3}$$

The first sum represents the Debye microwave relaxation with resonance frequency ω_r. The second sum accounts for the infrared, visible and ultraviolet adsorption spectra, where the constant g_j describes the bandwidth of the adsorption frequency ω_j. Table IV lists the data used in the calculations presented here.

Appendix B. Drainage Rates in a Capillary

CYLINDRICAL CAPILLARY

Consider flow of oil and water in a cylindrical capillary perpendicular to the plane of Figure 2. The flow is directed downwards along the z direction. We assume that $r \gg t$. Define a coordinate x that is zero on the solid surface. $x = w$ is the water/oil interface and $x = w + t$ is the oil/gas interface. For laminar flow near a solid surface, v_z is the only nonzero component of the velocity. We will assume that the water film does not contribute to the flow, that the oil is draining freely under gravity, that there is a free boundary condition at the gas/oil interface and a no flow boundary at the solid (or, equivalently, the water/oil interface). The flow velocity in the oil film is:

$$v_z = \frac{(x - w)}{\mu_o} \rho_o g \left[t - \frac{(x - w)}{2} \right], \quad t + w \geq x \geq w. \tag{B.1}$$

The total oil flux through an annular layer of thickness t and circumference $2\pi r$ is found from integrating Equation (B.1):

$$Q = \frac{2\pi \rho_o g r t^3}{3\mu_o}. \tag{B.2}$$

The time t_d taken to drain a volume V is Q/V.

TABLE IV. Table of constants used to compute the dielectric response. The data for water comes from Parsegian (1981) and that for octane and quartz from Hough and White (1980)

	C_r	$\omega_r(10^{11}$ rad/s)	
Water microwave	74.8	1.05	
	$\omega_j(10^{14}$ rad/s)	f_j	$g_j(10^{14}$ rad/s)
infrared	0.319	1.429	0.228
	1.048	0.735	0.577
	1.398	0.154	0.425
	3.038	0.143	0.380
	6.38	0.079	0.851
ultraviolet	126.1	0.0392	7.75
	151.9	0.057	13.5
	173.2	0.0923	23.4
	197.5	0.156	31.1
	226.3	0.152	45.0
	281.0	0.271	95.1
n-Octane			
infrared	5.54	0.023	0
ultraviolet	186.3	0.925	0
Quartz			
infrared	2.093	1.93	0
ultraviolet	203.2	1.359	0

SQUARE CAPILLARY

The fluid configuration is shown in Figures 4 and 5. We assume that the water and gas are stationary and take the water pressure at $z = 0$ to be zero. We further assume that the radius of curvature of the gas/oil interface is $r/2$. The water and gas pressures as a function of height are

$$P_w = -(h_1 + h_2 + z)\rho_w g, \tag{B.3}$$

$$P_g = -h_1\rho_w g - h_2\rho_o g + \frac{2\gamma_{ow}}{r} + \frac{2\gamma_{go}}{r}, \tag{B.4}$$

assuming that the gas density is negligible. The pressure difference between the oil and water phases is

$$P_o - P_w = \frac{\gamma_{ow}}{r_{ow}} = P_o + (h_1 + h_2 + z)\rho_w g. \tag{B.5}$$

Equation (B.5) is used to find r_{ow} and then the thickness of the oil layer $t = (\sqrt{2}-1)(r_{go} - r_{ow})$ is

$$t = \frac{(\sqrt{2}-1)r}{2}\left(\frac{f-x}{f}\right), \tag{B.6}$$

$$f = P_o + (h_1 + h_2 + z)\rho_w g, \tag{B.7}$$

$$x = \frac{2\gamma_{ow}}{r}, \tag{B.8}$$

We can write a Darcy like equation for the oil flow:

$$Q = \frac{crt^3}{\mu_o}\left(\frac{\partial P_o}{\partial z} + \rho_o g\right), \tag{B.9}$$

where Q is the flow rate. In analogy to the analysis of a cylindrical film (B.2), the conductance of the oil layer is written as crt^3, which assumes that we have film drainage with a free gas/oil interface. c is a dimensionless geometrical factor that is also affected by the boundary condition at the oil/water interface and is thus influenced by both the oil and water viscosities. Substituting (B.6) into (B.9) yields:

$$\frac{\partial f}{\partial z} = (\rho_w - \rho_o)g + m\left(\frac{f}{f-x}\right)^3, \tag{B.10}$$

where

$$m = \frac{8Q\mu_o}{(\sqrt{2}-1)^3 cr^4}. \tag{B.11}$$

We can not solve (B.10) analytically. However, the second term in (B.10) is always larger than the first, and so to a good approximation, if the density difference between the oil and water phases is small, we may write:

$$\left(\frac{f-x}{f}\right)^3 \frac{\partial f}{\partial z} = m \tag{B.12}$$

which has the solution

$$mh = (f_h - f_0) - 3x\ln\left(\frac{f_h}{f_0}\right) - 3x^2\left(\frac{1}{f_h} - \frac{1}{f_0}\right) + \frac{x^3}{2}\left(\frac{1}{f_h^2} - \frac{1}{f_0^2}\right), \tag{B.13}$$

where f_h and f_0 are the values of f at $z = h$ and $z = 0$ respectively: $f_0 = x + (\rho_w - \rho_o)gh_2$ and $f_h = x + (\rho_w - \rho_o)gh_2 + \rho_w gh$. We can find the flow rate from (B.11):

$$Q = \frac{(\sqrt{2}-1)^3 r^4 c}{8\mu_o} m, \tag{B.14}$$

where m is given by (B.13). Q increases with h, reaching an asymptotic value where $m = \rho_w g \approx \rho_o g$ when h is very large. The time t_d taken to drain a ganglion of volume V is Q/V with Q given by Equations (B.13) and (B.14).

Acknowledgements

Financial support for this work was provided by the Stanford University Petroleum Research Institute Gas Injection Affiliates Program.

References

1. Adamson, A. A.: 1990, *Physical Chemistry of Surfaces*, Wiley, New York.
2. Bear, J.: 1972, *Dynamics of Fluids in Porous Media*, Dover, New York.
3. Blunt, M. J., Fenwick, D. H., and Zhou, D.: 1994, What determines residual oil saturation in three phase flow?, SPE 27816, *Improved Oil Recovery Symposium*, Tulsa, OK.
4. Brooks, R. H., and Corey, A. T.: 1966, Properties of porous media affecting fluid flow, *J. Irrigation Drainage Division of the ASCE* **92**, 61–88.
5. Chatzis, I., Kantzas, A., and Dullien, F. A. L.: 1988, On the investigation of gravity-assisted inert gas injection using micromodels, long Berea Sandstone cores, and computer-assisted tomography, SPE 18284, *63rd Ann. Tech. Conf. Exhib. of the SPE*, Houston, TX.
6. Childs, W. H. J.: 1949, *Physical Constants*, Methuen, London.
7. Corey, A. T.: 1986, *Mechanics of Immiscible Fluids in Porous Media*, Water Resources Publications, Littleton, CO, 253 pp.
8. Derjaguin, B. V., and Kussakov, M. M.: 1939, Anomalous properties of thin polymolecular films V, *Acta Physicochim. URSS* **10**, 25.
9. Dong, M., Dullien, F. A. L., and Chatzis, I.: 1994, Imbibition of oil in film form over water present in edges of capillaries with an angular cross-section, *J. Colloid Interface Sci.* in press.
10. Dullien, F. A. L.: 1992, *Porous Media, Fluid Transport and Pore Structure*, Academic Press, San Diego.
11. Dumoré, J. M., and Schols, R. S.: 1974, Drainage capillary-pressure functions and the influence of connate water, *Soc. Petroleum Engrs. J.* (November), 437–444.
12. Dzyaloshinskii, I. E., Lifshitz, E. M., and Pitaevskii, L. P.: 1961, The general theory of van der Waals forces, *Adv. Phys.* **10**, 165–209.
13. Genuchten, M. T. V.: 1980, A closed-form equation for predicting the hydraulic conductivity of unsaturated soils, *Soil Sci. Soc. Amer. J.* **44**, 892–898.
14. Gibbs, J. W.: 1928, *The Collected Works of J. Willard Gibbs*, Longmans, Green, New York.
15. Hirasaki, G. J.: 1988, Wettability: fundamentals and surface forces, SPE 17367, *SPE/DOE Enhanced Oil Recovery Symp.* Tulsa, Oklahoma.
16. Hirasaki, G. J.: 1993, Structural interactions in the wetting and spreading of van der Waals fluids, *J. Adhesion Sci. Technol.* **7**(3), 285–322.
17. Hough, D. B., and White, L. R.: 1980, The calculation of Hamaker constants from Lifshitz theory with applications to wetting phenomena, *Adv. Colloid and Interface Sci.* **14**, 3–41.
18. Kalaydjian, F. J.-M.: 1992, Performance and analysis of three-phase capillary pressure curves for drainage and imbibition in porous media, SPE 24878, *67th Ann. Tech. Conf. Exhib. SPE*, Washington, DC.
19. Kalaydjian, F. J.-M., Moulu, J.-C., Vizika, O., and Munkerud, P.K.: 1993, Three-phase flow in water-wet porous media: determination of gas/oil relative permeabilities under various spreading conditions, SPE 26671, *68th Ann. Tech. Conf. Exhib. SPE*, Houston, TX.
20. Kantzas, A., Chatzis, I., and Dullien, F. A. L.: 1988a, Enhanced oil recovery by inert gas injection, SPE 17379, *SPE/DOE Symp. Enhanced Oil Recovery*, Tulsa, OK.
21. Kantzas, A., Chatzis, I., and Dullien, F. A. L.: 1988b, Mechanisms of capillary displacement of residual oil by gravity-assisted inert gas injection, SPE 17506, *SPE Rocky Mountain Regional Meeting*, Casper, WY.

22. Lenhard, R. J., and Parker, J. C.: 1990, Estimation of free hydrocarbon volume from fluid levels in monitoring wells, *Ground Water* **28**(1), (January–February), 57–67.
23. Leverett, M. C.: 1941, Capillary behavior in porous solids, *Trans AIME*, **142**, 152–169.
24. Lifshitz, E. M.: 1955, *J. Exp. Theor. Phys.* **29**, 94.
25. Mahanty, J., and Ninham, B. W.: 1976, *Dispersion Forces*, Academic Press, New York.
26. McBride, J. F., Simmons, C. S., and Cary, J. W.: 1992, Interfacial spreading effects on one-dimensional organic liquid imbibition in water-wetted porous media, *J. Contaminant Hydrol.* **11**, 1–25.
27. Mohanty, K. K., Davis, T., and Scriven, L. E.: 1987, Physics of oil entrapment in water-wet rock, *SPERE*, **2**(1), 113–128.
28. Morrow, N. R., Chatzis, I., and Taber, J. J.: 1988, Entrapment and mobilization of residual oil in bead packs, *SPE Reservoir Engrg.* (August), 927–934.
29. Muskat, M.: 1949, *Physical Principles of Oil Production*, McGraw Hill, Boston.
30. Oren, P. E., Billiotte, J., and Pinczewski, W. V.: 1992, Mobilization of waterflood residual oil by gas injection for water-wet conditions, *Soc. Petroleum Engrs. Formation Evaluation*, (March), 70–78.
31. Oren, P. E., and Pinczewski, W. V.: 1991, The effect of film flow on the mobilization of waterflood residual oil by immiscible gas flooding, SPE, *6th Eur. IOR Symp.*, Stavanger, Norway.
32. Oren, P. E., and Pinczewski, W. V.: 1992, The effect of wettability and spreading coefficients on the recovery of waterflood residual oil by miscible gas flooding, SPE, *67th Ann. Tech. Conf. Exhib. SPE*, Washington, DC.
33. Parker, J. C., Lenhard, R. J., and Kuppusamy, T.: 1987, A parametric model for constitutive properties governing multiphase flow in porous media, *Water Resour. Res.* **23**, 618–624.
34. Parsegian, V. A., and Weiss, G. H.: 1981, Spectroscopic parameters for computation of van der Waals forces, *J. Colloid Interface Sci.* **81**, 285–289.
35. Roof, J. G.: 1970, Snapoff of oil droplets in water-wet pores, *Soc. Petroleum Engrs. J.* **10**, 85.
36. Rowlinson, J. S., and Widom, B.: 1989, *Molecular Theory of Capillarity*, Clarendon, Oxford.
37. Soll, W. E., Celia, M. A., and Wilson, J. L.: 1993, Micromodel studies of three-fluid porous media systems: pore-scale processes relating to capillary pressure-saturation relationships, *Water Resources Res.* **29**(9), 2963–2974.
38. Vizika, O.: 1993, Effect of the spreading coefficient on the efficiency of oil recovery with gravity drainage, SPE, *Symp. on Enhanced Oil Recovery*, presented before the Division of Petroleum Chemistry, Inc., Denver, CO.
39. Vizika, O., and Lombard, J.-M.: 1994, Wettability and spreading: two key parameters in oil recovery with three-phase gravity drainage, SPE, *69th Ann. Tech. Conf. Exhib.*, New Orleans, LA.

Transport in Porous Media **20:** 105–133, 1995.

Fluid Distribution and Pore-Scale Displacement Mechanisms in Drainage Dominated Three-Phase Flow

P. E. ØREN[1] and W. V. PINCZEWSKI[2]
[1]*Statoil Research Centre, N-7005 Trondheim, Norway*
[2]*Australian Petroleum Cooperative Research Centre, University of New South Wales, Sydney, NSW 2052, Australia*

(Received: May 1994)

Abstract. This paper presents a precise description of the fluid distribution and pore-scale displacement mechanisms for three-phase flow under strongly wetting conditions when the displacing fluid is a nonwetting phase. It is shown that on the pore-scale the fluids may adopt one of three basic configurations depending on the values of the three interfacial tensions and the wetting preference of the solid. The nature of the three-phase displacement mechanisms is determined by the pore-scale fluid distribution. The displacing phase may advance by two basic mechanisms; a double drainage mechanism involving all three phases – a three-phase displacement – or, a direct drainage mechanism – a two-phase displacement. The three-phase displacement mechanism is described by a simple generalisation of two-phase flow mechanisms. The basic displacement mechanisms are incorporated into a numerical percolation-type network model which is used to compute phase recoveries for three-phase displacements. Computed recoveries are shown to be in good agreement with those determined experimentally. The model may therefore provide a basis for modelling three-phase flows in actual porous media.

Key words: Three-phase flow, fluid distribution, pore-scale displacement mechanisms, spreading coefficient, film flow.

1. Introduction

Three-phase flow in porous media is of interest in many areas of engineering and science. Important industrial examples arise in enhanced oil recovery processes where gases are injected into reservoirs containing oil and water, in groundwater flows where the introduction of organic liquids due to surface spills and to seepage from subsurface storage tanks pose serious problems to groundwater quality, and in radionuclide migration from deposits of nuclear waste.

Macroscopic multi-phase flow in porous media is usually described in terms of Darcy's law and measured saturation-dependent relationships for phase-relative permeabilities and capillary pressures. For three-phase flow these relationships are difficult to determine experimentally, and three-phase behaviour is almost always estimated from two-phase saturation-dependent data on the basis of empirical models. The models used are similar to the model first proposed by Leverett

(1941) and later extended by others (e.g., Stone, 1970, 1973; Parker and Lenhard, 1990). The empirical nature of these models (Kalaydjian, 1992) constitutes a major deficiency in the present theory for three-phase flow in porous media.

In principle, it is not necessary to actually measure two-phase or three-phase relative permeability and capillary pressure data since it is possible to determine these by appropriately averaging the equations describing the physical processes occurring on the microscopic or pore scale. This approach requires a detailed understanding of displacement mechanisms on the pore scale and a complete description of the morphology of the pore space. The precedure has been successfully applied to two-phase flow in simple or idealised porous media using pore-scale physics identified in micromodel experiments (Lenormand *et al.*, 1983) with the morphology of the pore space represented by a topologically equivalent numerical network (Blunt and King, 1991; Blunt *et al.*, 1992; Billiotte *et al.*, 1993). The references cited provide good examples of the current state of the art for two-phase flow. Although the extension of these techniques to real porous media is complicated by the difficulty of describing the complex nature of the pore space, network models have played an important role in improving our understanding of multi-phase flow and will prove to be valuable aids in interpreting and extending difficult and tedious laboratory measurements, particularly in three-phase flow.

The application of network modelling techniques to three-phase flow is considerably less developed than for two-phase flow. This is because little is currently understood of the pore-scale physics for three-phase displacements. Previous work by the authors (Øren and Pinczewski, 1991, 1994; Øren *et al.*, 1992, 1994), concerned with the mobilisation and recovery of waterflood residual oil by immiscible gas flooding, suggests that it may now be possible to provide a sufficiently complete description of the pore-scale displacement processes to allow realistic network models to be developed for three-phase flow, particularly for strongly wetting conditions when the invading fluid is a nonwetting phase such as gas (of specific interest to the oil industry) or an organic phase introduced into a groundwater system (groundwater quality).

The purpose of the present paper is to present a precise description of the fluid distribution and the pore-scale displacement mechanisms for drainage dominated three-phase flow under strongly wetting conditions. It is shown that for strongly wetting conditions, three-phase displacements may be completely described by a simple generalisation of previously described two-phase flow mechanisms in which the role of wetting in two-phase flow is replaced by spreading for three-phase flow. This would be of considerable interest to those contemplating the extension of two-phase network modelling techniques to three-phase flow.

2. Basic Concepts

The three-phase flows of interest in the present study all involve the injection of a *nonwetting phase* (gas) into a porous medium which is originally saturated with two liquid phases (oil and water), one of which strongly wets the solid. For simplicity, we consider only the case when the wetting phase completely wets the solid. Mixed wet and intermediate wet systems are beyond the scope of the present discussion.

We consider that the porous medium may be represented by an equivalent network of interconnected pores in which larger pores (pore bodies) are connected by narrower pores (pore throats). This description has been shown to provide realistic models for an actual porous media (Ioannidis and Chatzis, 1993). General procedures for characterising porous media have been recently reviewed by Dullien (1991). Although detailed features of the network, such as pore body size distribution, pore body-to-throat aspect ratio, and pore body coordination number, are known to affect fluid distribution and flow (Diaz *et al.*, 1987), the underlying pore-scale physics remains the same.

The distribution of the fluids within the pore space determines the transport properties of the phases. The static equilibrium pore-scale distribution of three fluids in a porous medium is determined by a complex interaction involving the following physical phenomena:

1. *Wettability* – interactions between the fluids and the solid.
2. *Capillary pressure* – interactions between bulk fluids across curved interfaces.
3. *Spreading phenomena and three-phase contact lines* – interactions between the three fluid phases.

Designating the fluids as fluid-1, fluid-2, and fluid-3, with fluid-3 being the wetting phase, the fluids may distribute in one of the three ways depicted schematically in Figure 1.

2.1. WETTABILITY

When two fluids (fluid-1 and fluid-3) are in contact with a solid (denoted by s), the wetting preference of the solid for fluid-3 relative to fluid-1 is given by the spreading coefficient S, which is defined as,

$$S = \sigma_{s1} - \sigma_{s3} - \sigma_{13}, \tag{1}$$

where σ_{s1}, σ_{s3}, and σ_{13} are the interfacial tensions for the solid–fluid-1, solid–fluid-3, and fluid-1–fluid-3 interfaces, respectively. When S is positive, fluid-3 wets the solid perfectly and the contact angle is zero. When the spreading coefficient is negative, the liquid only partially wets the solid. For this case the contact angle is given by Young's relation,

$$\cos\theta = 1 + \frac{S}{\sigma_{13}}. \tag{2}$$

For the present discussion, fluid-3 wets the solid perfectly and the solid surface is everywhere covered by a thin wetting film (see Figure 1).

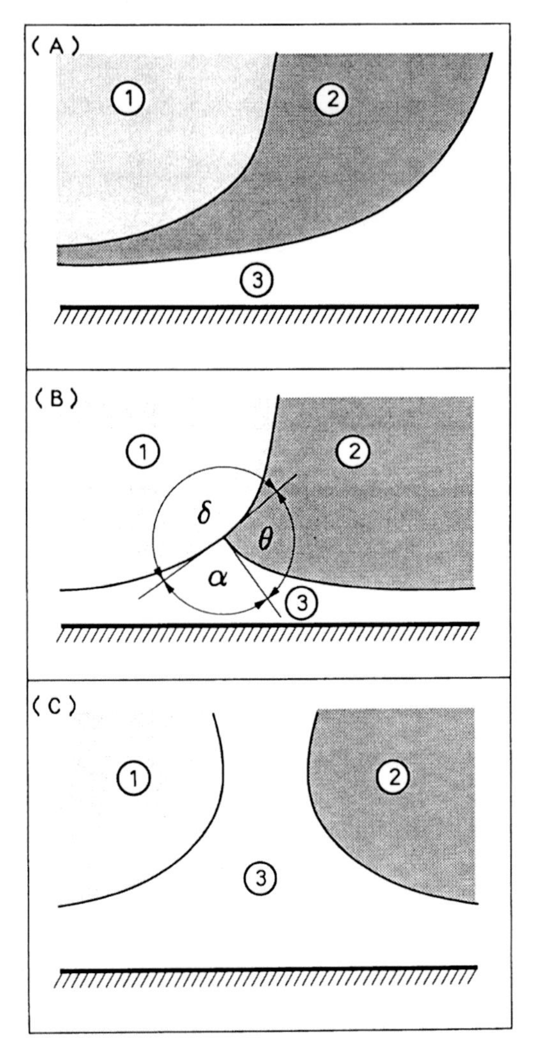

Fig. 1. Three-phase fluid interfaces for strongly wetting conditions.

2.2. CAPILLARY PRESSURE

When a nonwetting fluid (fluid-1 or fluid-2) is located in a pore body which is connected to a pore throat filled with wetting fluid (fluid-3), capillary forces prevent the nonwetting fluid from spontaneously entering the pore throat. The nonwetting fluid can only enter the pore throat when the pressure in the nonwetting fluid exceeds the pressure in the wetting fluid by a value equal to the threshold capillary pressure, P_c, given by

$$P_c = \sigma \left[\frac{1}{R_1} + \frac{1}{R_2} \right] = \frac{2\sigma}{r_t}, \tag{3}$$

where σ is the interfacial tension and R_1 and R_2 are the principal radii of curvature of the interface. The last term may be regarded as defining an effective radius of curvature, r_t, for the interface in the pore throat.

When the interface is initially located in a pore throat, there exists an excess capillary pressure, equal to the difference between the capillary pressure for the interface in the throat and that for the interface in the pore body, which causes the nonwetting fluid to spontaneously invade an adjacent pore body filled with wetting fluid. The excess capillary pressure, ΔP_c, is given by

$$\Delta P_c = 2\sigma \left[\frac{1}{r_t} - \frac{1}{r_b} \right], \qquad (4)$$

where r_t and r_b are the effective radii of the pore throat and pore body, respectively.

The local equilibrium distribution of two fluids in a porous medium, when one of the fluids completely wets the solid, is therefore determined by capillary pressure. As a result of capillary pressure, the nonwetting phase preferentially occupies pore bodies which may be interconnected by nonwetting phase filled pore throats, whilst the wetting phase preferentially occupies pore throats which may be interconnected by wetting phase filled pore bodies.

For the case of three fluids in a porous medium, there are two nonwetting phases (fluid-1 and fluid-2). In regions of the pore space occupied by the wetting phase and only one of the nonwetting phases, the distribution of fluids is similar to that of a two-phase system. However, in regions occupied by all three phases, one of the nonwetting fluids will have the greater excess capillary pressure as defined by Equation (4). This will be fluid-1 when,

$$\sigma_{13} > \sigma_{23}. \qquad (5)$$

When this condition is satisfied, we refer to fluid-1 as the *nonwetting phase* and fluid-2 as the *intermediate phase.*

2.3. SPREADING PHENOMENA AND THREE-PHASE CONTACT LINES

In parts of the pore space occupied by all three fluids, the equilibrium configuration of the phases may take one of the three forms shown schematically in Figure 1. In Figure 1(A), the intermediate phase, fluid-2, spontaneously spreads between fluids-1 and -3 to completely wet the wetting phase (fluid-3). In Figure 1(B), fluid-2 only partially wets fluid-3 forming a three-phase contact line characterised by the three contact angles α, θ, and δ. In Figure 1(C), fluid-2 neither spreads nor forms a contact line and the nonwetting and intermediate phases are completely contained within the wetting phase. The actual configuration adopted by a particular three-phase system will depend on the wetting preference of fluid-3 and the condition governing the existence of stable contact angles.

The wetting preference of fluid-3 is given by the spreading coefficient, S_{23}, defined as,

$$S_{23} = \sigma_{31} - \sigma_{32} - \sigma_{12}, \tag{6}$$

where σ_{31}, σ_{32}, and σ_{12} are the interfacial tensions for the fluid-3–fluid-1, fluid-3–fluid-2, and fluid-1–fluid-2 interfaces, respectively. When S_{23} is positive, fluid-2 spreads between fluids-1 and -3 to completely enclose fluid-1 and the fluids adopt the configuration shown in Figure 1(A).

When the spreading coefficient is negative, fluid-2 does not spread and the fluids may adopt either of the configurations shown in Figure 1(B and C). The actual configuration adopted depends on the condition governing the existence of a stable three-phase contact line. If a three-phase contact line exists, three forces due to interfacial tensions act at this line. In the absence of line tension, a force balance between the three phases at equilibrium requires that the vectorial equation,

$$\boldsymbol{\sigma}_{13} + \boldsymbol{\sigma}_{23} + \boldsymbol{\sigma}_{12} = 0, \tag{7}$$

must be satisfied. The above equation, known as the *law of Neumann's triangle* (Dullien, 1979), can be used to compute the contact angles α, θ, and δ,

$$\cos\alpha = \frac{\sigma_{12}^2 - \sigma_{23}^2 - \sigma_{13}^2}{2\sigma_{13}\sigma_{23}}, \tag{8}$$

$$\cos\theta = \frac{\sigma_{13}^2 - \sigma_{23}^2 - \sigma_{12}^2}{2\sigma_{12}\sigma_{23}}, \tag{9}$$

$$\cos\delta = \frac{\sigma_{23}^2 - \sigma_{12}^2 - \sigma_{13}^2}{2\sigma_{12}\sigma_{13}}. \tag{10}$$

Although the postulate of the existence of a line tension is open to conjecture (del Cerro and Jameson, 1978), the overall effect on the contact angles predicted by the above equations is likely to be small (Pujado and Scriven, 1972; del Cerro and Jameson, 1978). The condition for the existence of a stable three-phase contact line may therefore be expressed as,

$$S_{23} < 0 \tag{11}$$

$$\frac{\sigma_{13}}{|(\sigma_{12} - \sigma_{23})|} > 1. \tag{12}$$

When the above inequalities are fulfilled, the fluids form a stable three-phase contact line and the distribution of the phases is as shown in Figure 1(B). If the inequalities are not fulfilled, and the spreading coefficient S_{23} is negative, there is

TABLE I. Interfacial tensions for the air–Soltrol–water

	σ_{wg}	σ_{og}	σ_{wo}	S_{ow}
Water	73.3	23.0	32.6	+17.7
Water/isobutanol	33.1	20.9	20.3	−8.1

neither a contact line nor fluid-2 spreading and the fluids adopt the configuration shown in Figure 1(C).

Equations (5–12) may be used to estimate the pore-scale fluid distribution for three-phase systems in porous media from a knowledge of the three interfacial tensions and the wetting preference of the solid. We demonstrate this below by considering each of the systems used in the three-phase micromodel displacement experiments described previously.

3. Observed Pore-Scale Fluid Configurations

The three-phase gas–oil–water displacement experiments described by Øren and Pinczewski (1991, 1994) and Øren *et al.* (1992, 1994) were carried out in two-dimensional glass micromodels under strongly *water wet* and strongly *oil wet* conditions with fluids forming *positive* and *negative spreading systems*. The spreading coefficient refers to oil spreading between water and gas,

$$S_{ow} = \sigma_{wg} - \sigma_{wo} - \sigma_{og}. \tag{13}$$

The fluids used in the displacements were air, Soltrol-130, and distilled water. The measured interfacial tensions are given in Table I and show that the air–oil–water system has a positive spreading coefficient ($S_{ow} = +17.7$). The addition of a small quantity of isobutanol to the water results in a negative spreading coefficient ($S_{ow} = -8.1$).

3.1. POSITIVE SPREADING SYSTEM – WATER WET CONDITIONS

A photograph of the three-phase fluid distribution in the micromodel is shown in Figure 2. Water is the wetting fluid (fluid-3 = *water*). From Equation (5), gas is the nonwetting fluid (fluid-1 = *gas*) since $\sigma_{gw} > \sigma_{ow}$. Oil is therefore the intermediate fluid (fluid-2 = *oil*). From Table I we have that

$$S_{23} = \sigma_{wg} - \sigma_{wo} - \sigma_{go} = +17.7.$$

As the spreading coefficient is positive, the fluids will adopt the configuration shown in Figure 1(A). The predicted existence of oil-spreading films is consistent with the experimental observations for this system.

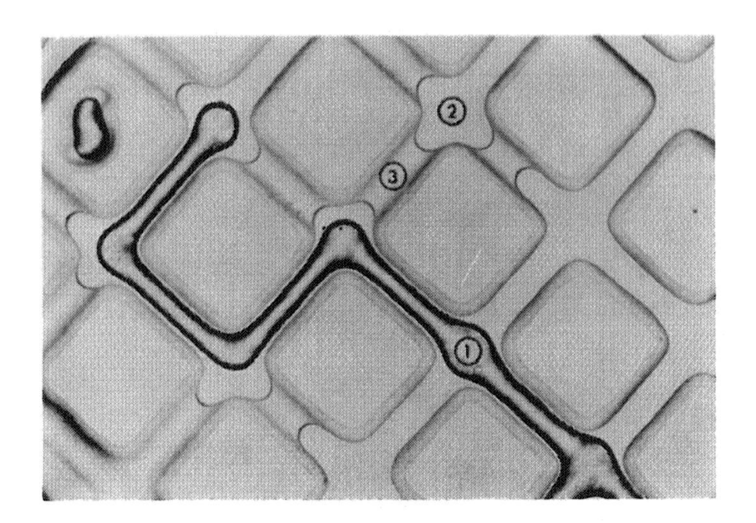

Fig. 2. Three-phase fluid distribution in glass micromodel for strongly water wet conditions and a positive spreading coefficient (gas – 1, oil – 2, water – 3).

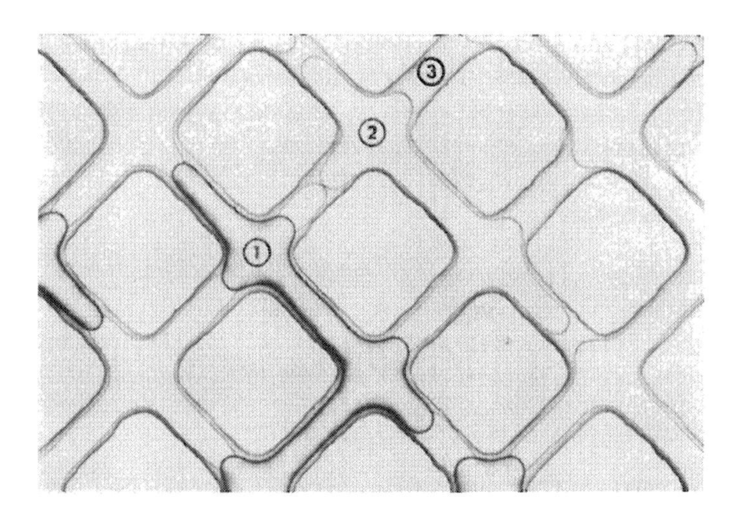

Fig. 3. Three-phase fluid distribution in glass micromodel for strongly water wet conditions and a negative spreading coefficient (gas – 1, oil – 2, water – 3).

3.2. NEGATIVE SPREADING SYSTEM – WATER WET CONDITIONS

A photograph of the three-phase fluid distribution for this system is shown in Figure 3. Water is the wetting fluid (fluid-3 = *water*). From Equation (5), gas is the nonwetting fluid (fluid-1 = *gas*) since $\sigma_{gw} > \sigma_{ow}$. Oil is therefore the intermediate fluid (fluid-2 = *oil*). From Table I we have that,

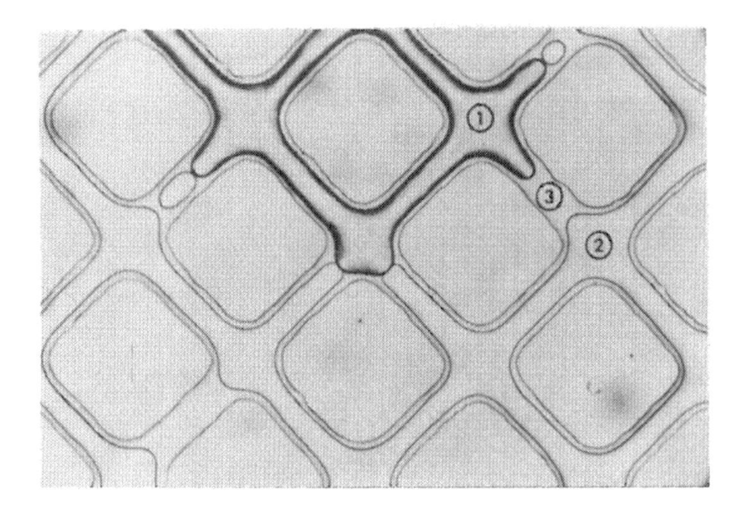

Fig. 4. Three-phase fluid distribution in glass micromodel for strongly oil wet conditions and a negative spreading coefficient (gas – 1, water – 2, oil – 3).

$$S_{23} = \sigma_{wg} - \sigma_{wo} - \sigma_{go} = -8.1$$

Since the spreading coefficient is negative, the fluids may adopt either of the configurations shown in Figure 1(B or C). We therefore test for the existence of a stable three-phase contact line (Equation 12),

$$\frac{\sigma_{gw}}{|(\sigma_{go} - \sigma_{ow})|} = 55.2 > 1$$

Since the inequality is fulfilled, the fluids adopt the configuration shown in Figure 1(B) with contact angles (Equations (8–10) $\alpha = 143°$, $\theta = 73°$, and $\delta = 144°$. The predicted existence of a three-phase contact line with the above contact angles is consistent with the experimental observations for this system.

3.3. NEGATIVE SPREADING SYSTEM – OIL WET CONDITIONS

A photograph of the three-phase fluid distribution for this system is shown in Figure 4. Oil is the wetting fluid (fluid-3 = *oil*). From Equation (5), gas is the non-wetting fluid (fluid-1 = *gas*) since $\sigma_{go} > \sigma_{wo}$. Water is therefore the intermediate fluid (fluid-2 = *water*). From Table I we have that,

$$S_{23} = \sigma_{og} - \sigma_{ow} - \sigma_{gw} = -32.5$$

Since the spreading coefficient is negative, the fluids may adopt either of the configurations shown in Figure 1(B or C). We therefore test for the existence of a stable three-phase contact line (Equation (12)),

$$\frac{\sigma_{go}}{|(\sigma_{gw} - \sigma_{wo})|} = 1.63 > 1$$

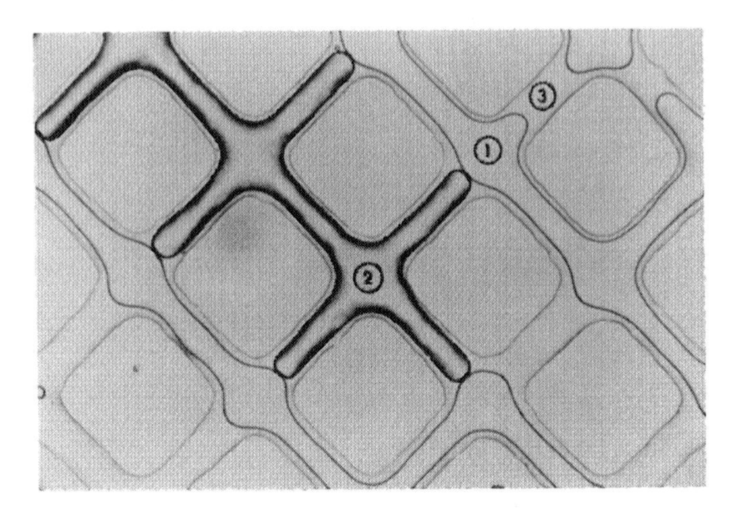

Fig. 5. Three-phase fluid distribution in glass micromodel for strongly oil wet conditions and a positive spreading coefficient (water – 1, gas – 2, oil – 3).

Since the inequality is fulfilled, the fluids adopt the configuration shown in Figure 1(B) with contact angles (Equations 8–10) $\alpha = 73°$, $\theta = 143°$, and $\delta = 144°$. The predicted existence of a three-phase contact line with the above contact angles is consistent with the experimental observations for this system.

3.4. POSITIVE SPREADING SYSTEM – OIL WET CONDITIONS

A photograph of the three-phase fluid distribution for this system is shown in Figure 5. Oil is the wetting fluid (fluid-3 = *oil*). From Equation (5), water is the nonwetting fluid (fluid-1 = *water*) since $\sigma_{wo} > \sigma_{go}$. Gas is therefore the intermediate fluid (fluid-2 = *gas*). From Table I we have that,

$$S_{23} = \sigma_{ow} - \sigma_{og} - \sigma_{wg} = -63.7$$

Since the spreading coefficient is negative, the fluids may adopt either of the configurations shown in Figure 1(B or C). We therefore test for the existence of a stable three-phase contact line (Equation 12),

$$\frac{\sigma_{wo}}{|(\sigma_{wg} - \sigma_{go})|} = 0.65 < 1$$

Since the inequality is not fulfilled, the fluids adopt the configuration shown in Figure 1(C). The predicted absence of a contact line and the complete wetting of both water and gas by oil is consistent with the experimental observations for this system.

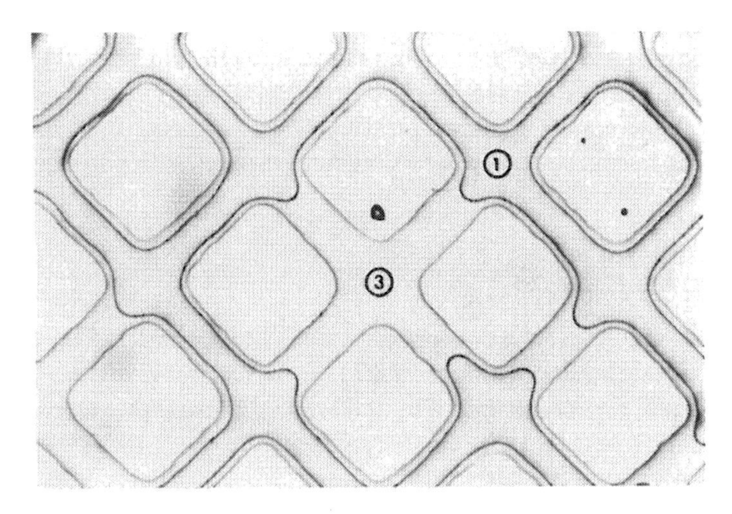

Fig. 6. Two-phase fluid distribution in glass micromodel for strongly water wet conditions (gas – 1, water – 3).

4. Static Three-Phase Fluid Distribution

The overall static distribution of three fluids in a glass micromodel for strongly wetting conditions is determined by the equilibrium pore-scale fluid configurations discussed above. The implications of these distributions on the connectivity of the phases are important in determining the nature of three-phase displacement processes and these are discussed below. Comparisons are made with results from previously reported studies for two-phase flow to highlight the similarities between two-phase and three-phase flow. The similarities between the three-phase fluid distributions and those for two-phase systems are clearly seen by comparing Figures 2–5 with Figure 6.

4.1. WETTING PHASE – FLUID-3

The distribution of the wetting phase is similar to that previously observed in two-phase studies (see Figures 2–6). For the case of complete wetting, the wetting phase is hydraulically connected throughout the porous network and it may either,

(i) completely saturate regions of interconnected pore bodies and throats (regions of high wetting phase saturation), or
(ii) exist as a thin wetting film on the surface of the pore space, in corners of pores with angular cross-sections and in microscopic grooves (regions of low wetting phase saturation).

The hydraulic continuity of the wetting phase, clearly observable in glass micromodels, has been confirmed in studies using actual porous rocks. Barci *et al.* (1985) showed that at low injection rates into sandstones pre-wet with wetting fluid, the wetting-phase saturation increases throughout the core without forming a distinct front, thus indicating transport of the wetting phase through thin films. Furthermore, Dullien *et al.* (1986) found no lower bound for the wetting-phase saturation in Berea cores with increasing capillary pressure, thus indicating that the wetting phase maintains hydraulic continuity throughout the rock down to very low saturations.

4.2. NON-WETTING PHASE –FLUID-1

A comparison between Figure 6 for two-phase systems and Figures 2–5 for three-phase systems shows that the configuration of the non-wetting phase in three-phase flow is similar to that in two-phase flow. The non-wetting phase is prevented from spontaneously entering a pore throat by capillary forces. As a result, the nonwetting phase preferentially occupies the larger pore bodies.

Unlike the wetting phase, which is always continuous, the nonwetting phase may be discontinuous. In regions where the nonwetting phase is continuous, it fills pore bodies and interconnecting pore throats. In regions where it is discontinuous, some of the pores may be completely filled with wetting or intermediate phase.

Although we speak of the nonwetting phase as filling a pore body or pore throat, it is understood that all the pores always contain a thin film of wetting fluid (as discussed above) and, in some cases (see discussion below), pores may also contain a thin spreading film of the intermediate phase (see Figure 2). These films are usually sufficiently thin for the effective curvature of the nonwetting phase interface to be approximately determined by the dimensions of the pore space.

Nonwetting phase may become disconnected during either drainage or imbibition displacements (Mohanty and Salter, 1982). Both types of displacements result in the movement of nonwetting phase through a pore throat. When the pore body to pore throat radius ratio is sufficiently large ($r_b/r_t > 3$, approximately), a capillary driven instability causes the nonwetting fluid in the pore throat to choke-off or snap-off (Mohanty and Salter, 1982; Chatzis and Dullien, 1983). This is a bond-breaking mechanism which leads to disconnection of the nonwetting phase.

4.3. INTERMEDIATE PHASE – FLUID-2

The intermediate phase is also a nonwetting phase, and its distribution is regions of the pore space occupied predominantly by the intermediate phase is similar to that for the nonwetting phase. Again, as for the nonwetting phase, the intermediate phase may become disconnected during either drainage or imbibition displacements by choke-off or snap-off events in pore throats.

In regions of the pore space occupied by both intermediate and nonwetting phases, the distribution of the phases is as shown in Figures 2–5. For all cases, capillary pressure determines that the nonwetting fluid preferentially occupies pore bodies with the intermediate phase located in adjoining pore throats.

Positive Spreading Coefficient – Figure 2

When the spreading coefficient S_{23} is positive, the intermediate fluid spreads between the wetting and nonwetting fluids to form a thin film which everywhere separates the nonwetting phase from the wetting phase. Since the nonwetting fluid preferentially occupies pore bodies, the transition between bulk intermediate phase and the thin spreading film occurs at the entrances to pore throats.

As the intermediate phase spreading film is continuous, it may be expected to play an important role in hydraulically connecting the intermediate phase, similar to that of the wetting film in connecting the wetting phase. The presence of spreading films allows,

(i) all disconnected intermediate fluid contacted by the advancing nonwetting and intermediate fluids to be reconnected, and
(ii) mobility of the intermediate fluid down to very low saturations.

Spreading films may therefore be anticipated to have an important effect on both the mobility and residual saturation of the intermediate phase in three-phase displacements.

Three-phase Contact Line – Figures 3 and 4

When the intermediate phase only partially wets the nonwetting phase, the three fluids form a three-phase contact line. Since the nonwetting phase preferentially occupies pore bodies, the three-phase contact line is located at the entrances to pore throats.

No Three-phase Contact Line and $S_{23} < 0$ – Figure 5

When the intermediate phase does not spread and there is no contact line, the intermediate and nonwetting fluids are separated by a thin film of wetting fluid, with the nonwetting phase occupying pore bodies and the intermediate fluid in adjoining pore throats.

The thin film separating the two nonwetting phases is established in the following manner. As a result of capillary forces, the intermediate fluid moves into the diverging part of the throat at the entrance to the pore body. The wetting fluid film separating the intermediate and nonwetting fluids drains under the influence of this pressure until the film thickness becomes sufficiently small (about 1000Å) for the film to become stabilised by a positive disjoining pressure resulting from a combination of London–van der Waals, electrostatic, and structural forces. The overall result is that at equilibrium, the nonwetting and intermediate fluids are separated by a thin film of wetting fluid located at the entrances to pore throats.

The physical situation is very similar to that which exists for the case where the three phases meet in a contact line and the connectivity of the phases is also similar for the two cases. The two cases, in fact, are very difficult to distinguish experimentally (see Figures 4 and 5).

5. Drainage Displacement Mechanisms in Three-Phase Flow

Assuming that throats are of uniform cross-section and that the displacement is sufficiently slow for viscous losses associated with the movement of an interface through a throat to be negligible, the condition for an interface to enter a throat from an adjoining pore body, displace fluid from the throat and reach the diverging pore space at the entry to the neighbouring pore body, is determined by capillary pressure. This forms the basis for the invasion percolation network models previously used to describe drainage displacements in two-phase flow (Wilkinson and Willemsen, 1983; Lenormand *et al.*, 1983; Blunt *et al.*, 1992; Billiotte *et al.*, 1993). In the following we extend these concepts to three-phase flows where a nonwetting fluid (fluid-1 or fluid-2) is injected into a strongly wetted (fluid-3) porous medium containing all three phases. The pore-scale displacement mechanisms are described for each of the three basic fluid configurations shown in Figure 1(A, B, and C). In the context of network modelling we refer to pore throats as links and pore bodies as nodes in the network.

5.1. CONFIGURATION A

This fluid configuration corresponds to the case of Figure 1(A). A fluid-2 spreading film everywhere separates fluid-1 and fluid-3. A number of examples of displacements involving this configuration are sketched in Figures 7–10. The injected phase, fluid-1, may advance by one of two basic displacement mechanisms:

(i) a *double drainage* mechanism – a three-phase displacement, or
(ii) a *modified direct drainage* mechanism – similar to a two-phase displacement.

Double drainage mechanism. Fluid-1 displaces fluid-2 by a double drainage mechanism in which a fluid 1–2 interface movement is always associated with a corresponding fluid 2–3 interface movement. A schematic of this mechanism is depicted in Figure 7. If we define the capillary pressure for this displacement as $P_c(= P_1 - P_3)$, the threshold capillary pressure for the double drainage mechanism may be written as,

$$P_c = \sigma_{12} C_{L1} + \sigma_{23} C_{L2} \tag{14}$$

where σ_{12} and σ_{23} are the fluid 1–2 and fluid 2–3 interfacial tensions and C_{L1} and C_{L2} are the effective curvatures of the interfaces in links (throats) $L1$ and $L2$. For strongly wetting conditions, we may approximate the effective curvatures by,

$$C_{L1} = \frac{2}{r_{L1}} \tag{15}$$

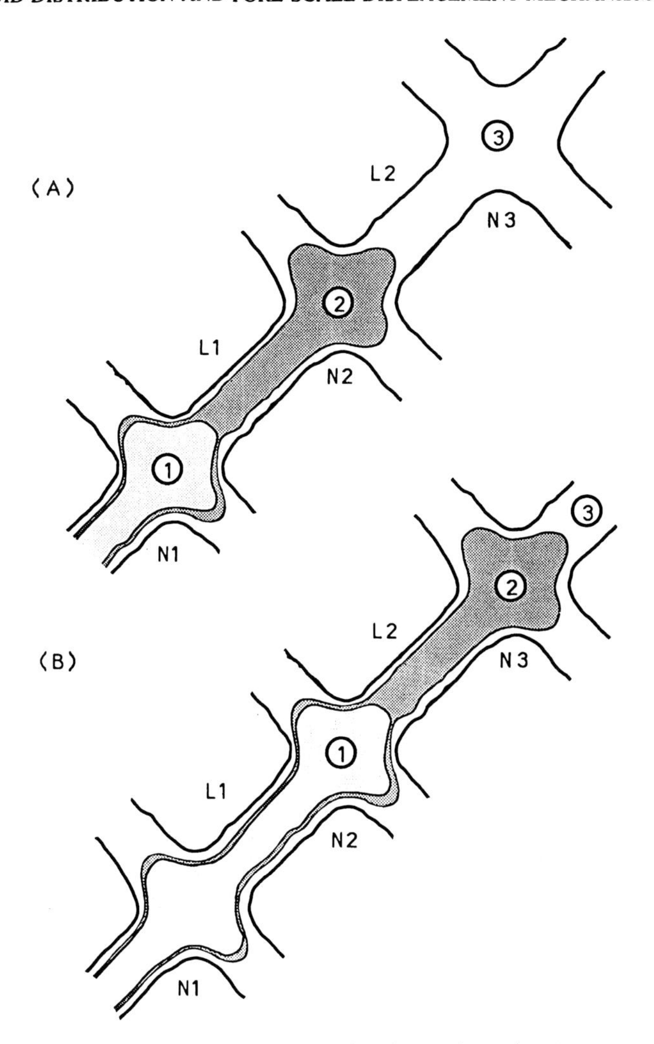

Fig. 7. Double drainage mechanism for configuration A.

where r_{L1} is the effective radius for link $L1$.

If fluid-2 is continuous to the outlet (the largest pore in the system), $C_{L2} \approx 0$ and,

$$P_c = \sigma_{12} C_{L1}. \tag{16}$$

For the case shown in Figure 7, the second drainage event – the interface involved in the fluid 2–3 displacement – is in close proximity to the first drianage event – the fluid 1–2 displacement. However, the presence of a continuous fluid-2 spreading film allows many possible sites for the second drainage event. These include:

(i) all the fluid 2–3 interfaces directly associated with the invaded fluid-2 blob, and

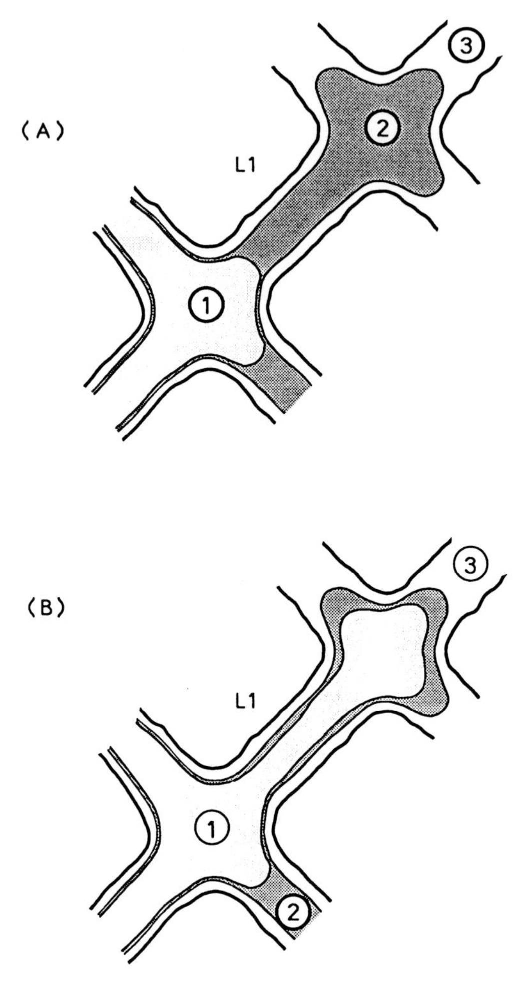

Fig. 8. Double drainage mechanism for configuration A with the displaced phase-2 flowing through spreading films.

(ii) all the fluid 2–3 interfaces along the fluid-2 spreading film separating fluid-1 and fluid-3 (all of phase-2 contacted by fluid-1).

An example of this is shown in Figure 8 where the displaced fluid-2 flows through films to the site of the second drainage event. The capillary pressure for this displacement is also given by Equation (14); however, C_{L2} neither needs to be an adjacent throat nor a throat directly associated with the fluid-2 blob being invaded, but may be any throat on the entire fluid 2–3 interface.

The double drainage mechanism may result in coalescence and reconnection of residual fluid-2. Figure 9 shows an example of this process. Here, the second drainage event results in fluid-2 in node $N2$ invading node $N3$ which already contains fluid-2. As the advancing fluid 2–3 interface moves along link $L2$, fluid-3

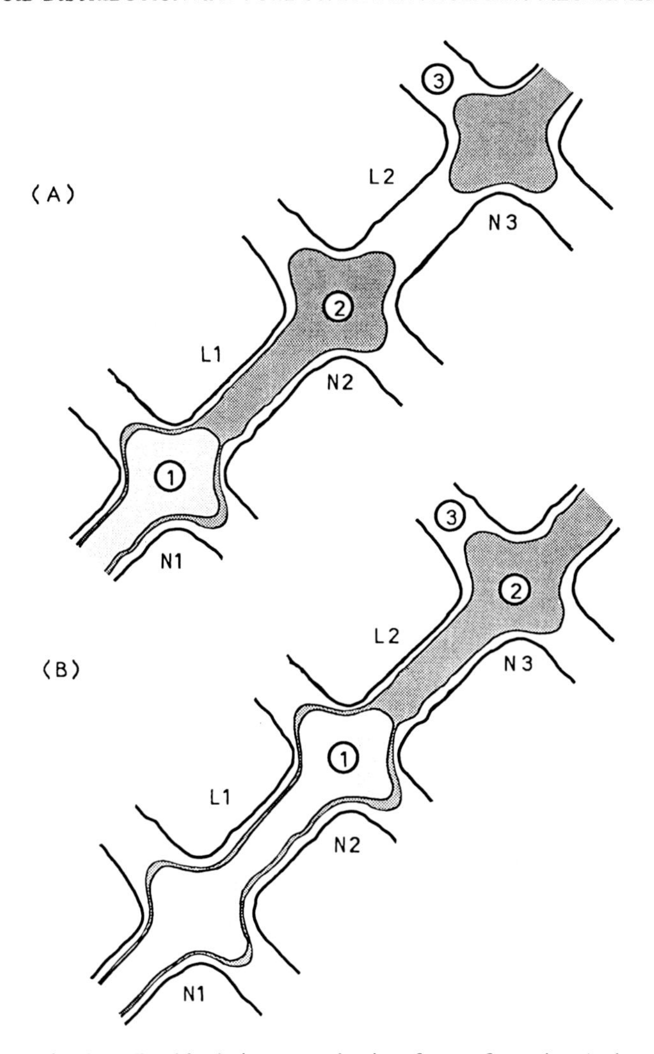

Fig. 9. Double drainage mechanism for configuration A showing blob reconnection as a result of coalescence.

in the invaded link is drained by flow through the wetting film until the advancing fluid 2–3 interface in node $N3$ is separated from fluid-2 by a thin film of wetting fluid. The wetting film drains under the action of capillary forces until it thins to a thickness where intermolecular forces become important. Provided that the intermolecular forces are attractive, the wetting film continues to thin and eventually ruptures, resulting in coalescence and reconnection of fluid-2 (Figure 9B).

Modified direct drainage mechanism. This displacement mechanism is depicted in Figure 10 and involves fluid-1, covered by a thin spreading film, displacing fluid-3. The mechanism is really a special case of the previously described double drainage mechanism in which both the first and second drainage events occur in the same

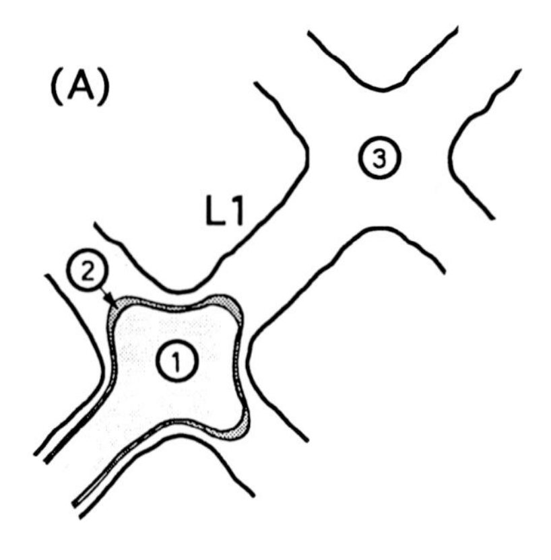

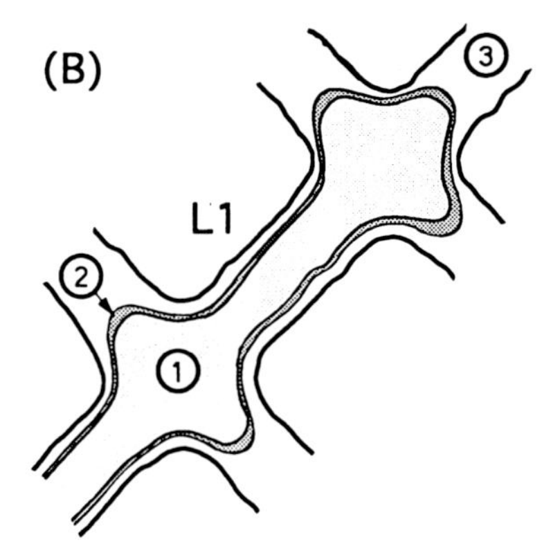

Fig. 10. Modified direct drainage mechanism for configuration A.

part of the pore space. The threshold capillary pressure for the displacement is thus given by,

$$P_c = (\sigma_{12} + \sigma_{23})C_{L1} \tag{17}$$

which is simply Equation (14) written with $C_{L1} = C_{L2}$.

5.2. CONFIGURATION B

A schematic of this fluid configuration is shown in Figure 1(B). The main differences between this fluid configuration and configuration A are the absence of intermediate phase spreading films and the presence of a three-phase contact line. An example of a displacement process for this fluid configuration is sketched in Figure 11. The injected fluid-1 may advance by one of two basic displacement mechanisms:

(i) a *double drainage* mechanism – a three-phase displacement, or

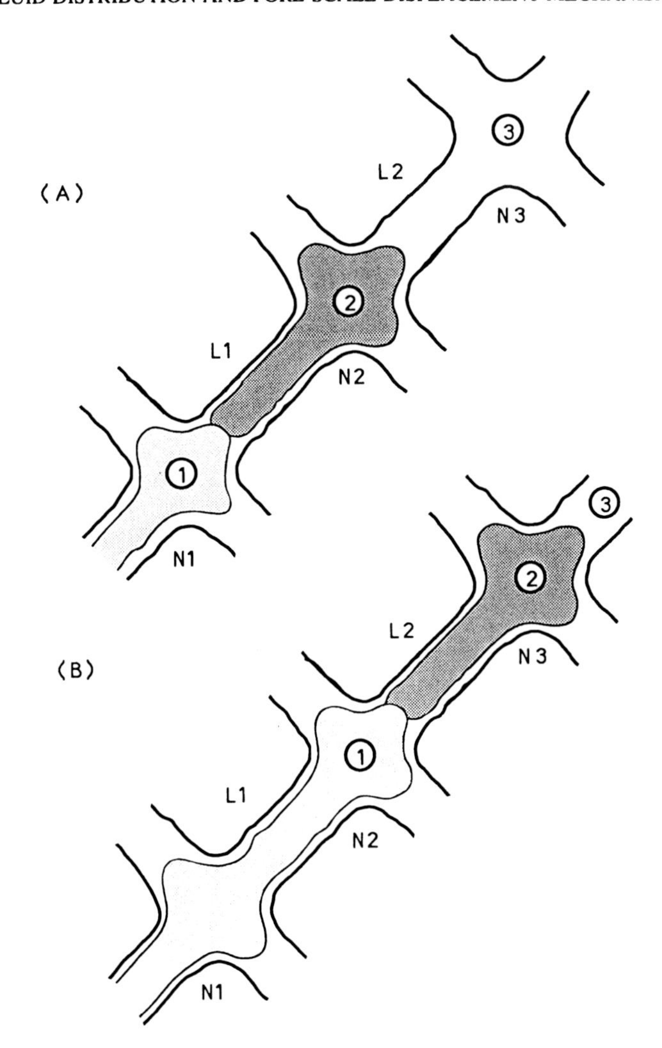

Fig. 11. Double drainage mechanism for configuration B.

(ii) a *direct drainage* mechanism – a two-phase displacement.

Double drainage mechanism. Except for the motion of the three-phase contact line at the site of the first drainage event, the double drainage mechanism shown in Figure 11 is similar to that for fluid configuration A when there are no spreading films. Neglecting the deformation of the wetting phase at the three-phase contact line, and in the absence of line tension, Equations (8–10) may be approximated as (Wallace and Schurch, 1988),

$$\cos\theta = \frac{\sigma_{13} - \sigma_{23}}{\sigma_{12}} = 1 + \frac{S_{23}}{\sigma_{12}}, \tag{18}$$

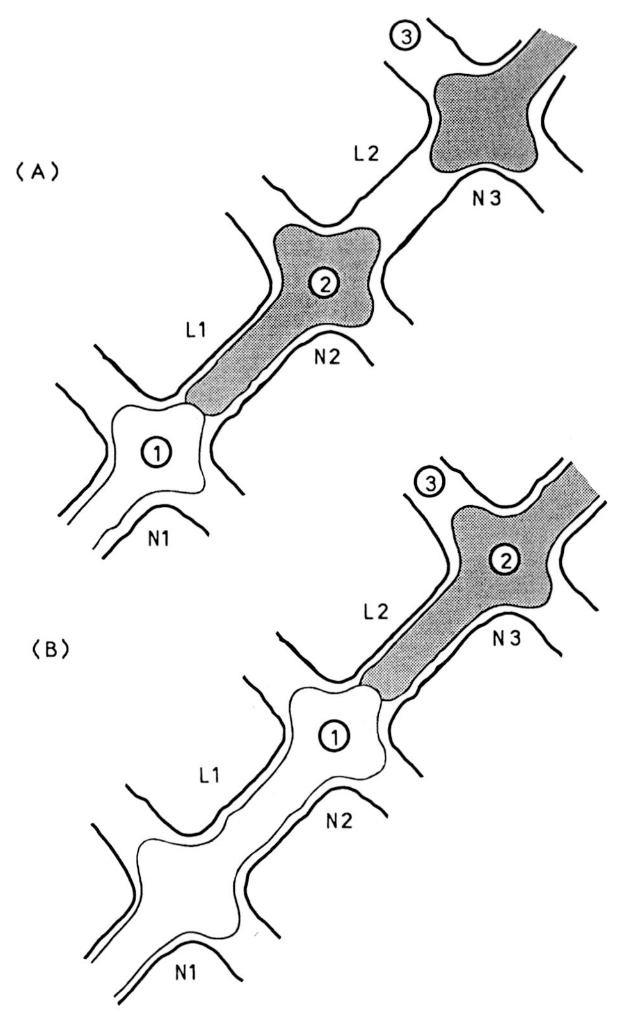

Fig. 12. Double drainage mechanism for configuration B showing blob reconnection as a result of coalescence.

where θ is the fluid 1–2 contact angle on the fluid-3 substrate. The threshold capillary pressure for the double drainage mechanism may then be written as,

$$P_c = \cos\theta\sigma_{12}C_{L1} + \sigma_{23}C_{L2}. \tag{19}$$

Another example of the double drainage mechanism is shown in Figure 12. Here, node $N3$ is occupied by fluid-2 rather than fluid-3 as in Figure 11. In this case, fluid-3 is displaced from link $L2$ until the advancing fluid 2–3 interface reaches the entrance of the fluid-2 filled node. As before, the wetting film separating the advancing fluid-2 in link $L2$ and fluid-2 in node $N3$ drains and ruptures, leading to coalescence in the node and reconnection of fluid-2.

Direct drainage mechanism. In regions of the pore space occupied only by fluid-1 and the wetting fluid (fluid-3), fluid-1 displaces fluid-3 by a direct drainage mechanism identical to that for two-phase flow. The capillary pressure for the direct drainage mechanism is given by,

$$P_c = \sigma_{13} C_{L1}. \tag{20}$$

5.3. CONFIGURATION C

For this fluid configuration the non-wetting and intermediate fluids are completely contained within the wetting fluid (Figure 1C). The main difference between this configuration and fluid configuration B is that the three-phase contact line for the latter is replaced by a thin film of fluid-3 in the case of the former. Both the location and extent of the thin film are similar to those for the contact line in fluid configuration B. The displacement mechanisms for both configurations are therefore similar.

As for configuration B, the injected fluid-1 may advance by one of two basic displacement mechanisms:

(i) a *double drainage* mechanism – a three-phase displacement, or
(ii) a *direct drainage* mechanism – a two-phase displacement.

Double drainage mechanism. A schematic of the double drainage mechanism for fluid configuration C is shown in Figure 13. The process is similar to that for configuration B shown in Figure 11. The threshold capillary pressure for the double drainage mechanism is given by,

$$P_c = (\sigma_{13} - \sigma_{23}) C_{L1} + \sigma_{23} C_{L2}. \tag{21}$$

If fluid-2 is continuous to the outlet ($C_{L2} \approx 0$), the above equation reduces to,

$$P_c = (\sigma_{13} - \sigma_{23}) C_{L1}. \tag{22}$$

As for fluid configuration B, the double drainage process may result in coalescence and reconnection of fluid-2 blobs.

Direct drainage mechanism. In regions of the pore space occupied only by fluid-1 and the wetting fluid (fluid-3), fluid-1 displaces fluid-3 by a direct two-phase drainage displacement similar to that described for fluid configuration B. The threshold capillary pressure for the direct drainage mechanism is thus given by Equation (20).

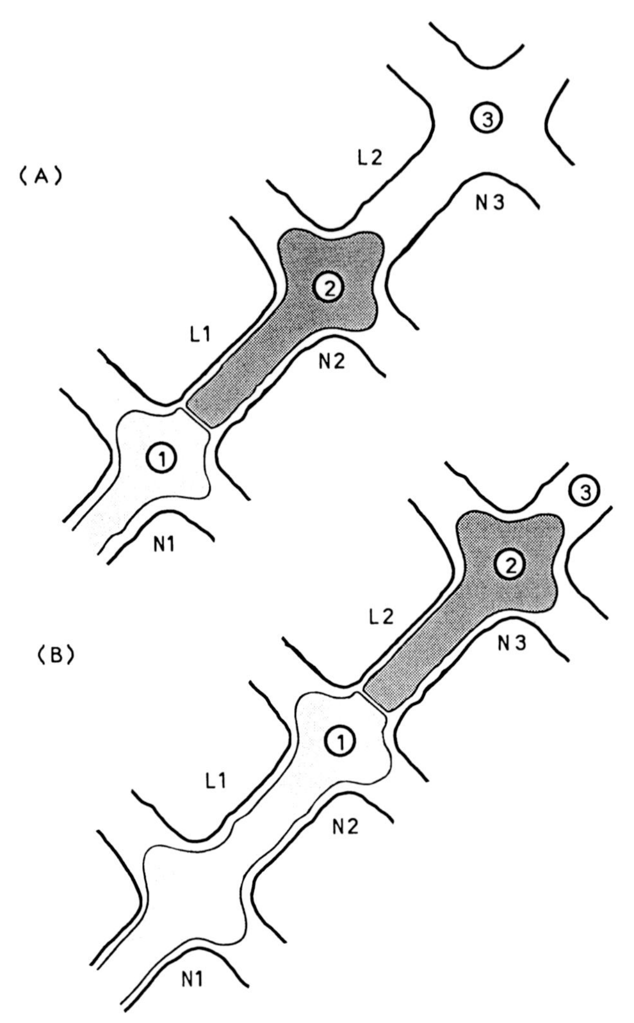

Fig. 13. Double drainage mechanism for configuration C.

5.4. FLUID CONFIGURATIONS B AND C WHEN FLUID-2 IS THE INJECTED PHASE

In the experiments reported by Øren and Pinczewski (1994) an interesting situation arises for the positive spreading system under oil wet conditions (Section 3.4). Here, the injected fluid is fluid-2 – the intermediate fluid – with the fluids distributed as for configuration C. A similar situation may arise for configuration B.

The main difference in the displacement mechanisms when fluid-2 is the injected phase is that the first step in the double displacement mechanism now involves fluid-2, which is initially located in a pore throat or link, invading a pore body or node filled with the non-wetting phase, fluid-1. This is an imbibition process rather than a drainage process.

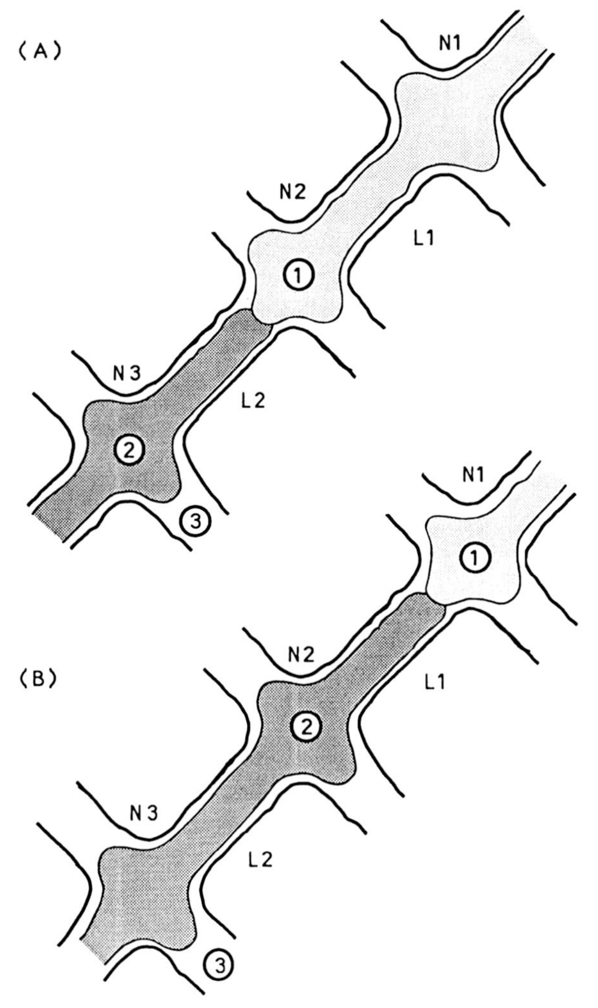

Fig. 14. Imbibition–drainage mechanism for configuration B when fluid-2 is the injected phase.

As before, the displacement mechanisms for fluid configurations B and C are similar. The injected fluid-2 may advance by one of two basic displacement mechanisms:

(i) an *imbibition-drainage* mechanism – a three-phase displacement, or
(ii) a *direct drainage* mechanism – a two-phase displacement.

Imbibition–drainage mechanism. Figure 14 shows a three-phase displacement for configuration B when fluid-2 is the injected phase. Since $\sigma_{23} < \sigma_{13}$, injected fluid-2 does not spontaneously invade fluid-1 occupied network nodes. Defining the capillary pressure for this displacement as $P_c(= P_2 - P_3)$, the threshold capillary pressure for the three-phase displacement is given by

$$P_c = -\cos\theta\sigma_{12}C_{N2} + \sigma_{13}C_L \tag{23}$$

where C_{N2} is the effective curvature of the fluid-2 invaded node $N2$ and C_L is the effective curvature of the fluid-1 invaded link L. Since the capillary pressure for the invasion of node $N2$ is negative, the invasion is a *forced imbibition* process and the two-step displacement is therefore an *imbibition–drainage* mechanism.

A three-phase displacement for configuration C when fluid-2 is the injected phase is shown in the photographic sequence of Figure 15(a–f). At the commencement of the displacement, fluid-2 is everywhere located at the entrances to fluid-1 filled nodes (Figure 15a). When the capillary pressure has increased sufficiently to overcome the threshold pressure, fluid-2 starts to invade the fluid-1 filled node (Figure 15b). As fluid-2 invades the node, thin bridges of fluid-1 are formed at the connecting link entrances (Figure 15b, d). The fluid-1 bridges thin and eventually rupture, resulting in disconnection and trapping of a small fluid-1 blob in the adjoining link (Figure 15e).

This three-phase displacement is similar to that described previously for configuration B (Figure 14). The threshold capillary pressure for the displacement is given by,

$$P_c = (\sigma_{23} - \sigma_{13})C_{N1} + \sigma_{13}C_{L2}. \tag{24}$$

As for configuration B, the invasion of node $N1$ is a forced imbibition process and the three-phase displacement is an imbibition–drainage mechanism. We note that for two-phase displacements, C_{N1} is directly related to the physical dimensions of the node being invaded (Lenormand *et al.*, 1983). This is not necessarily the case for three-phase displacements and is certainly not so for the experiment shown in Figure 15 where the effective curvature of the interfaces in the invaded node is very small. The shapes of the displacing interfaces in the node are similar to those in the flat contact region formed when two like but immisicible drops or bubbles approach each other. If C_{N1} is small, Equation (24) may be written as,

$$P_c = \sigma_{13}C_{L2}. \tag{25}$$

Similar to the double drainage mechanism, the imbibition–drainage process may result in coalescence and reconnection of fluid-1 blobs.

Direct drainage mechanism. In regions of the pore space occupied only by fluid-2 and the wetting phase, fluid-2 displaces fluid-3 by a direct two-phase drainage mechanism. The threshold capillary pressure for the direct drainage mechanism is given by,

$$P_c = \sigma_{23}C_{L1}. \tag{26}$$

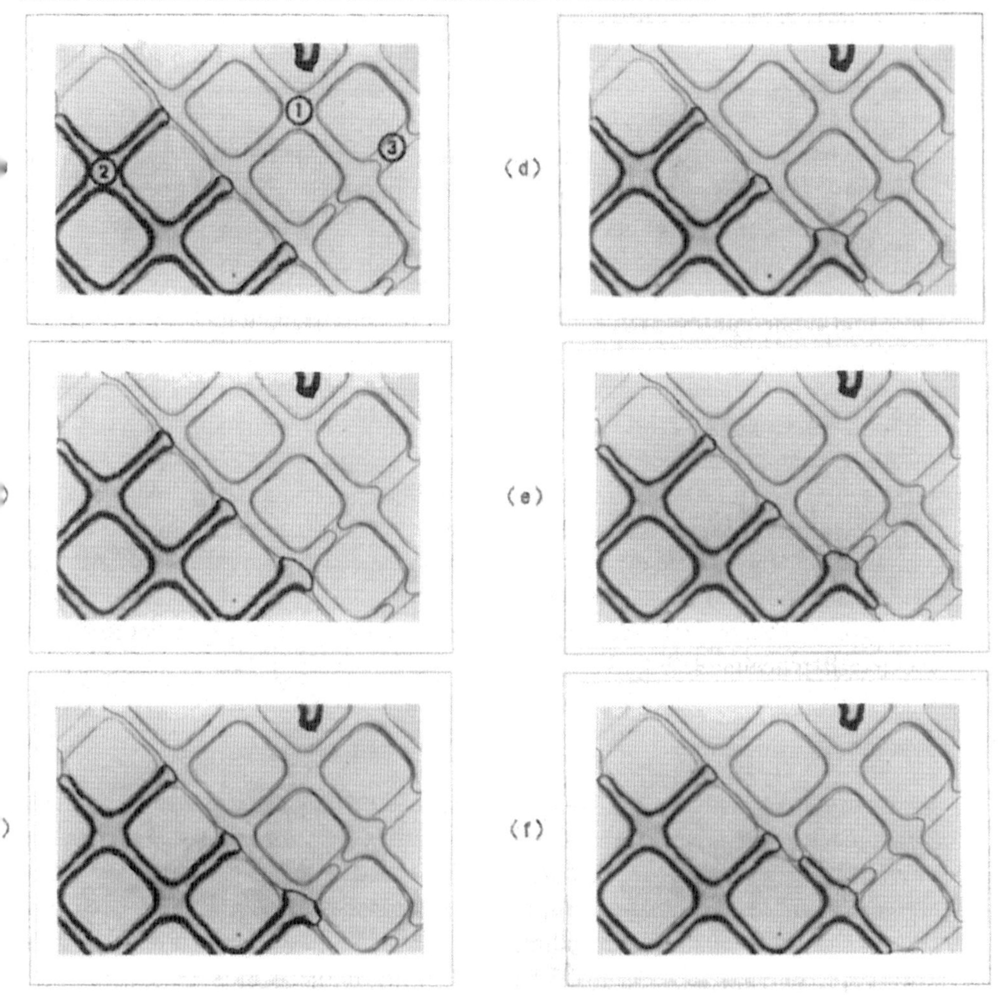

Fig. 15. Imbibition–drainage mechanism for configuration C when fluid-2 is the injected phase (water – 1, gas – 2, oil – 3).

5.5. BASIC DISPLACEMENTS

For drainage-dominated displacements when fluid-1 is the displacing phase, we may summarise the basic displacement mechanisms for three-phase flow by noting that for a displacement to proceed it is necessary to satisfy the following conditions:

(1) The displacing phase (fluid-1) must flow from the inlet to the site of the first displacement event.

(2) The displacing phase displaces the wetting phase (fluid-3) – a two-phase displacement – or the intermediate phase (fluid-2), the first displacement event of a double drainage process.

(3) The displaced wetting phase must flow to the outlet, or the displaced intermediate phase must flow to the site of the second displacement event of the double drainage process.
(4) In the case of a double drainage process, the displaced intermediate phase must displace wetting fluid – the second drainage event. If the site of this event is at the outlet, intermediate phase is produced.
(5) Wetting phase displaced in the second drainage event must flow to the outlet.

A similar set of conditions may be written for the case when fluid-2 is the displacing phase.

For the case of systems having a positive spreading coefficient S_{23}, both wetting fluid and intermediate fluid may be transported to and from displacement sites by flow through wetting and spreading films, respectively. These are generally higher resistance flow paths and therefore result in a considerable slowing of the overall displacement process. If the displacement is carried out with fluid-1 injected at a constant rate, the capillary pressure may increase during the course of the displacement to values exceeding the threshold pressure for the next invasion event. For these conditions a second invasion may commence prior to the completion of the first displacement.

6. Network Modelling

The above displacement sequences have recently been implemented into a rule-based percolation algorithm to simulate three-phase displacements under strongly wetting conditions both with and without intermediate phase spreading films. Details of the invasion algorithm, the phase-trapping rules and the approximate treatment of flow through spreading films used in the network model are given elsewhere (Øren *et al.*, 1994). The results from the network model simulations provide a test for the validity of some of the three-phase displacement mechanisms described above.

6.1. COMPARISON WITH EXPERIMENTS

Tertiary gas flood simulations were carried out for strongly water wet conditions with fluids forming positive and negative spreading systems (i.e. fluid configurations A and B). The interfacial tensions used in the simulations were the same as those used in the micromodel experiments (see Table I).

Figure 16 compares the experimentally observed and computed gas distributions for the positive spreading system at the later stages of the tertiary gas flood. Despite the complex nature of the three-phase displacements, simple capillary dominated invasion concepts provide a surprisingly accurate representation of the experiments. The displacement patterns for the gas in the experiments and in the simulations are qualitatively very similar. In the simulations, gas advances relatively uniformly

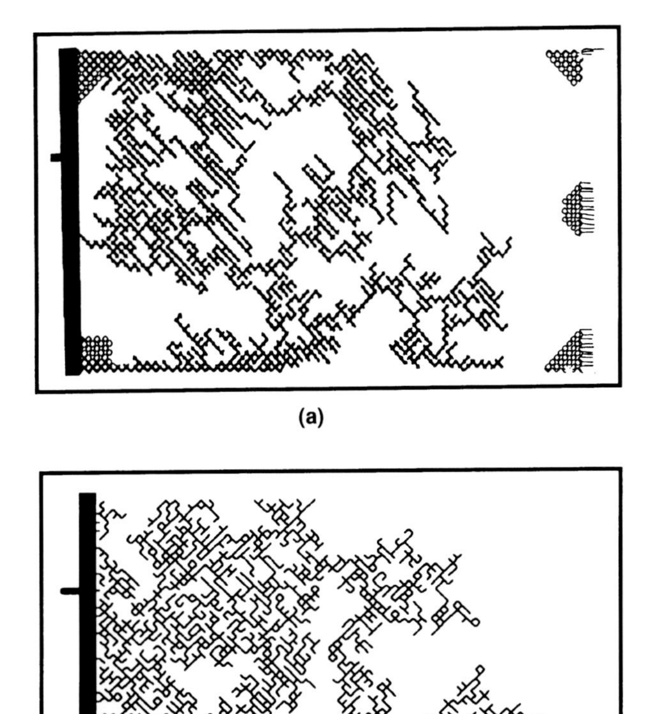

(a)

(b)

Fig. 16. (a) Experimental and (b) computed distribution of injected gas during the later stages of a tertiary displacement.

across the width of the network forming a distinct but highly dendritic gas front. This is in good agreement with the experimental observations.

Experimental and simulated oil recoveries were estimated from the number of pore bodies and throats containing oil before and after the gas flood. This information is summarised in Table II. Simulated oil recoveries are in close agreement with those observed experimentally (see Table II). The residual oil saturation for the negative spreading system was 31% for the simulations compared to 30% for the micromodel experiment. The corresponding residual oil saturations for the positive spreading system are 22% (simulated) and 24% (experimental).

The above results demonstrate that simple capillary invasion percolation concepts may be used to predict the behaviour of three-phase micromodel displacement experiments.

TABLE II. Experimental and simulated oil recovery data for water wet displacements

System	Status	Experiments			Simulations		
		Pores	Throats	$S_{or}(\%)$	Pores	Throats	$S_{or}(\%)$
$S_{ow} > 0$	Waterflood	2110	2508	37	2156	3089	41
$S_{ow} > 0$	Gas Flood	958	2121	24	981	1825	22
$S_{ow} < 0$	Waterflood	2133	2486	37	2156	3089	41
$S_{ow} < 0$	Gas Flood	992	2970	30	1002	3087	31

7. Conclusions

For conditions where one of the phases strongly wets a solid, the pore-scale distribution of three phases in a porous medium is uniquely determined by wettability, capillary pressure, and the spreading behaviour of the fluids.

On the pore-scale, three-phase displacements occur by a double drainage mechanism when the non-wetting phase is the displacing phase, and an imbibition–drainage mechanism when the displacing phase is the intermediate phase.

The three-phase displacement mechanism may be completely described by a simple generalisation of previously described two-phase flow mechanisms in which the role of wetting in two-phase flow is replaced by spreading for three-phase flow.

Three-phase displacement mechanisms may compete with two-phase displacement mechanisms. As for two-phase flow, the sequence of pore invasions is determined by pore threshold capillary pressures.

The fluid distributions and displacement mechanisms described above, combined with simple invasion percolation concepts modified to account for the presence of intermediate phase spreading films, provide an accurate description of complex three-phase flows in strongly wet glass micromodels. The three-phase displacement mechanisms reported in this paper may therefore provide a basis for modelling three-phase flows in actual porous media.

References

Bacri, J. C., Leygnac, C., and Salin, D.: 1983, Evidence of capillary hyperdiffusion in two-phase fluid flows, *J. Phys. Lett.* **46**, L467–472.

Billiotte, J. A., De Moegen, H., and Øren, P. E.: 1993, Experimental micromodeling and numerical simulation of gas/water injection/withdrawal cycles as applied to underground gas storage, *SPE Adv. Tech. Ser.* **1**(1), 133–139.

Blunt, M., King, M. J., and Scher, H.: 1992, Simulation and theory of two-phase flow in porous media, *Phys. Rev. A* **46**(12), 7680–7699.

Blunt, M. and King, P.: 1991, Relative permeabilities from two- and three-dimensional pore-scale network modelling, *Transport in Porous Media* **6**, 407–433.

del Cerro, M. C. G. and Jameson, G. J.: 1978, in J. F. Padday (ed) *Wetting, Spreading and Adhesion*, Academic Press, New York, pp. 61–82.

Chatzis, I. and Dullien, F. A. L.: 1983, Dynamic immiscible displacement mechanisms in pore doublets, *J. Coll. Interf. Sci.* **91**, 199–222.

Diaz, C. E., Chatzis, I. and Dullien, F. A. L.: 1987, Simulation of capillary pressure curves using bond correlated site percolation on a simple cubic network, *Transport in Porous Media* **2**, 215–240.
Dullien, F. A. L., Francis, S., Lai, Y., and Macdonald, I. F.: 1986, Hydraulic continuity of residual wetting phase in porous media, *J. Coll. Interf. Sci.* **109**(1), 201–218.
Dullien, F. A. L.: 1991, Characterization of porous media – Pore level, *Transport in Porous Media* **6**, 581–606.
Dullien, F. A. L.: 1979, *Porous Media Fluid Transport and Pore Structure*, Academic Press, New York.
Ioannides, M. A., and Chatzis, I.: 1993, Network modelling of pore structure and transport properties of porous media, *Chem. Eng. Sci.* **48**, 951–972.
Kalaydjian, F.: 1992, Performance and analysis of three-phase capillary pressure curves for drainage and imbibition in porous media, *paper SPE 24878, 67th Ann. Tech. Conf. Exhib. SPE*, Washington, DC, Oct. 4–7, 1992.
Lenormand, R., Zarcone, C., and Sarr, A.: 1983, Mechanism of the displacement of one fluid by another in a network of capillary ducts, *J. Fluid Mech.* **135**, 337–353.
Leverett, M. C.: Capillary behaviour in porous solids, *Trans. AIME Petrol. Eng. Div.* **142**, 152–169.
Li, D. and Neumann, A. W.: 1992, Equation of state for interfacial tensions of solid–liquid systems, *Adv. Colloid Int. Sci.* **39**, 299–345.
Mohanty, K. K. and Salter, S. J.: 1982, Multiphase flow in porous media – II. Pore level modelling, *paper SPE 11018, 57th SPE Tech. Conf.*, New Orleans, Sep. 26–29, 1982.
Øren, P. E. and Pinczewski, W. V.: 1991, The effect of film flow on the mobilization of waterflood residual oil by gas flooding, *Proc. 6th Eur. IOR-Symp.*, Stavanger, Norway, May 21–23, 1991, Vol.1. 705–716.
Øren, P. E., Billiotte, J. and Pinczewski, W. V.: 1992, Mobilization of waterflood residual oil by gas injection for water-wet conditions, *SPE Formation Evaluation* **7**(1), 70–78.
Øren, P. E. and Pinczewski, W. V.: 1994, Effect of wettability and spreading on the recovery of waterflood residual oil by immiscible gas flooding, *SPE Formation Evaluation* **9**(2), 149–156.
Øren, P. E., Billiotte, J. and Pinczewski W. V.: 1994, Pore-scale network modelling of waterflood residual oil recovery by immiscible gas flooding, *paper SPE/DOE 27814, 9th Symp. IOR*, Tulsa, OK, April 17–20, 1994.
Parker, J. C. and Lenhard, R. J.: 1990, Determining three-phase permeability-saturation-pressure relations from two-phase system measurements, *J. Petrol. Sci. Eng.* **4**, 57–65.
Pujado, P. R. and Scriven, L. E.: 1972, Sessile lenticular configurations: transactionally and rotationally symmetric lenses, *J. Coll. Interf. Sci.* **40**(1), 82–98.
Stone, H. L.: 1970, Probability model for estimating three-phase relative permeability, *J. Petrol. Technol.* **22**, 214–218.
Stone, H. L.: 1973, Estimation of three-phase relative permeability and residual oil data, *J. Canad. Petrol. Technol.* **12**(4), 53–61.
Wallace, J. A. and Schurch, S.: 1988, Line tension of a sessile drop on a fluid–fluid interface modified by a phospholipid monolayer. *J. Coll. Interf. Sci.* **124**, 452–461.
Wilkinson, D. and Willemsen, J. F.: 1983, Invasion percolation: a new form of percolation theory, *J. Phys. A* **16**, 3365–3376.

Transport in Porous Media **20:** 135–168, 1995.

Generalized Relative Permeability Coefficients during Steady-State Two-Phase Flow in Porous Media, and Correlation with the Flow Mechanisms

D. G. AVRAAM and A. C. PAYATAKES
Department of Chemical Engineering, University of Patras, and Institute of Chemical Engineering and High Temperature Chemical Processes, PO Box 1414, GR 26500 Patras, Greece

(Received: 7 June 1994; in final form: 14 November 1994)

Abstract. A parametric experimental investigation of the coupling effects during steady-state two-phase flow in porous media was carried out using a large model pore network of the chamber-and-throat type, etched in glass. The wetting phase saturation, S_1, the capillary number, Ca, and the viscosity ratio, κ, were changed systematically, whereas the wettability (contact angle θ_e), the coalescence factor Co, and the geometrical and topological parameters were kept constant. The fluid flow rate and the pressure drop were measured independently for each fluid. During each experiment, the pore-scale flow mechanisms were observed and videorecorded, and the mean water saturation was determined with image analysis. Conventional relative permeability, as well as generalized relative permeability coefficients (with the viscous coupling terms taken explicitly into account) were determined with a new method that is based on a B-spline functional representation combined with standard constrained optimization techniques. A simple relationship between the conventional relative permeabilities and the generalized relative permeability coefficients is established based on several experimental sets. The viscous coupling (off-diagonal) coefficients are found to be comparable in magnitude to the direct (diagonal) coefficients over board ranges of the flow parameter values. The off-diagonal coefficients (k_{rij}/μ_j) are found to be unequal, and this is explained by the fact that, in the class of flows under consideration, microscopic reversibility does not hold and thus the Onsager–Casimir reciprocal relation does not apply. The *coupling indices* are introduced here; they are defined so that the magnitude of each coupling index is the measure of the contribution of the coupling effects to the flow rate of the corresponding fluid. A correlation of the coupling indices with the underlying flow mechanisms and the pertinent flow parameters is established.

Key words: two-phase flow, relative permeabilities, ganglion dynamics, viscous coupling, coupling indices.

1. Nomenclature

Bo	=	bond number.
Ca	=	capillary number $= \mu_1 q_1/\gamma_{12} wl$.
Co	=	coalescence factor (effective probability of coalescence, given a collision between two ganglia in the porous medium).
C_k^i	=	parameters used in the functional representation of k_{ri} in terms of cubic B-splines, (Equation (10a)).
C_k^{ij}	=	parameters used in the functional representation of k_{rij} in terms of cubic B-splines, (Equation (10b)).

$e_{\mu\alpha}$ = residual for the μth experiment and the αth equation of the model, (Equation (6)).
$\mathbf{e}$ = vector of residuals, $e_{\mu\alpha}$.
k = absolute permeability.
k_{ri} = (conventional) relative permeability to fluid i.
k_{ri}^{o} = value of k_{ri} free from end and boundary effects.
k_{rij} = generalized relative permeability coefficients.
k_{rij}^{o} = value of k_{rij} free from end and boundary effects.
L = distance along which ΔP_1 and ΔP_2 are measured.
l = node-to-node distance of the pore network.
l_{α} = number of unknown parameters in the αth equation of the model.
N = number of cubic B-splines used to represent k_{ri} or k_{rij}, ((Equation (10a,b)).
n = number of experimental data.
q_i = flowrate of fluid i.
S_i = saturation of fluid i.
v_i = superficial velocity of fluid i.
$\mathbf{V}$ = covariance matrix of the true errors $\varepsilon_{\mu\alpha}$, for all experiments (μ) and equations (α) of the model.
$\mathbf{W}$ = weighing matrix, (Equation (7)).
w = width of the network.
$\mathbf{x}$ = vector of the independent variables, (Equation (5)).
$\mathbf{y}$ = vector of the dependent variables.

Greek Letters
γ_{12} = interfacial tension.
$\varepsilon_{\mu\alpha}$ = true error for the μth experiment and the αth equation.
ΔP_i = pressure drop (negative) in fluid i, along a distance L.
θ_e = equilibrium contact angle.
θ = vector of unknown parameters, (Equations (5) and (7)).
θ^* = value of θ that minimizes the objective function Φ, (Equation (7)).
$\hat{\theta}$ = true (but unknown) value of θ.
κ = μ_2/μ_1 = viscosity ratio.
μ_i = viscosity of fluid i.
$\sigma_{\mu\alpha}^2$ = variance of the error in the μth experiment and in the αth equation.
Φ = objective function, (Equation (7)).
$\Phi^{(1)}, \Phi^{(2)}$ = objective function for Model 1 and Model 2, respectively, (Equations (9a)–(b)).
χ_i = coupling index for fluid i.
χ_i^{o} = value of χ_i free from end and boundary effects.

Subscripts
1 water.
2 oil.

2. Introduction

Immiscible two-phase flow in porous media is encountered in many processes of great interest, such as oil recovery, agricultural irrigation, pollution of ground water aquifers by liquid wastes, etc. Such flows can have various modes: imbibition, drainage, steady-state, co-current, counter-current, etc. Here we are concerned with steady-state co-current flow of two immiscible fluids, water (short for wetting fluid; fluid 1), and oil (short for non-wetting fluid; fluid 2).

Recently, it has become clear that the interactions between the two fluids (viscous coupling effects) can be significant for a wide range of two-phase flow cases. However, there are still several important problems to be solved. Key among these are: (a) to develop a reliable and convenient method to obtain the values of all four generalized relative permeability coefficients from steady-state flow rate vs. pressure drop data, and (b) to correlate these coefficients with the pore-level flow mechanisms and, eventually, to understand the coupling effects in quantitative terms. In the present work, a multivariable nonlinear parameter estimation method is used to determine the generalized relative permeability coefficients, with the viscous coupling coefficients taken explicitly into account, during steady-state co-current two-phase flow experiments. In addition, the corresponding conventional relative permeabilities are also evaluated, from the same experimental data. The generalized coefficients are compared with the corresponding conventional relative permeabilities, and all coefficients are correlated with the flow parameters and the underlying flow mechanisms at the pore level, and the flow regimes at the macroscopic scale, in order to quantify the relative importance of the viscous coupling coefficients. A strikingly simple relationship between the generalized relative permeability coefficients and the conventional relative permeabilities, Equation (16), is deduced from theoretical considerations and several sets of experimental results. This relationship helps to explain the reason for which viscous coupling effects are so elusive, despite the fact that they usually are of the first magnitude: they exist embedded in the conventional relative permeabilities and can be extracted only through specially designed means.

Experimental studies of two-phase flow mechanisms are conveniently done in transparent model pore networks etched in glass (Wardlaw, 1982; Chatzis *et al.*, 1983; Lenormand *et al.*, 1983; Chen, 1986; Vizika and Payatakes, 1989; Ioannidis *et al.*, 1991). The experimental results used here have been taken from an experimental study of steady-state co-current two-phase flow in a large planar model pore network of the chamber-and-throat type, etched in glass (Avraam and Payatakes, 1995). In that study, the conventional relative permeabilities to both phases were measured directly using Darcy's law, and then correlated with the corresponding flow regimes. Furthermore, a comparative study of the flow phenomena in both planar and non-planar models of the same type has shown that the flow phenomena remain qualitatively the same, irrespective of the 2-D or 3-D topology (Avraam *et al.*, 1994). This provides justification for using large planar model pore networks. Evaluation of the generalized relative permeability coefficients was not attempted in either of the two aforementioned works for lack of a well established method. The difficulty encountered in such cases is that the four generalized relative permeability coefficients appear in two equations which relate two flow rates (q_1, q_2)

with two pressure drops (ΔP_1, ΔP_2). Hence, one steady-state experiment, at a given set of conditions (fixed S_1, Ca, κ, θ_e, Co, Bo etc.), does not provide sufficient information to determine the four coefficients. This difficulty has led many researchers to a quest for the design of experiments in which the coupling coefficient can be isolated (e.g., counter-current flow, etc). Such experiments can, indeed, prove the existence of strong coupling effects, but they can not be expected to yield the values of the coupling coefficients that pertain to steady-state co-current two-phase flow, given the large difference in the experimental conditions. Here, we develop a method which can determine these coefficients from a set of experiments of the same type (steady-state co-current two-phase flow) with rigorous statistical means. It is a parameter estimation method that obtains the functional dependence of the generalized coefficients in a global way by choosing the optimal functional representation of the unknown coefficients in order to match the experimental data with statistical rigor.

This method has the advantage that it uses only one set of experiments of the same type, which is both convenient and imperative. In this way, the experimental data are well controlled and correspond to the exact values of the parameters and to the flow mode and flow regimes which pertain to the flow system of interest. As will be seen below, the new method is simple and works very well.

The flow mechanisms and the macroscopic flow regimes affect the conventional relative permeabilities strongly (Avraam and Payatakes, 1995). In the past, the presumption has prevailed that the configuration of the fluids remains static under steady-state two-phase flow conditions. This, in turn, has led to a longstanding microconception concerning the mechanisms of the motion of the two phases. This assumption, which was first made by Richards (1931), presumes that a disconnected fluid cannot flow through the pore network, and, therefore, oil can flow only through connected pathways. On the other hand the oil (nonwetting fluid) is thought to be always disconnected below a certain saturation value (Honarpoor and Mahmood; 1988), in which case its permeability is thought to become nil. In Avraam and Payatakes (1994) it is shown that this dogma is not true (see also below), and in many actual situations of steady-state two-phase flow the disconnected parts of the fluids contribute substantially to the overall motion.

Avraam and Payatakes (1995) identified four main flow regimes: *large ganglion dynamics* (LGD), *small ganglion dynamics* (SGD), *drop traffic flow* (DTF) and *connected pathway flow* (CPF). Snapshots of such flows are given in Figure 1a–d. In the first two flow regimes, the motion of the oil is due to the dynamic process of ganglion motion, collision and coalescence, breakup, stranding and remobilization, leading to an overall dynamic equilibrium which was denoted as '*steady-state*' *ganglion dynamics*. The term 'steady-state' appears in quotation marks to show that the process is intrinsically a dynamic equilibrium of moving fluid parts, even though macroscopically it appears stable. Connected pathway flow for both fluids was observed only in the case of high flow rates. However, even in this case, elements of SGD and/or DTF were present in between the connected

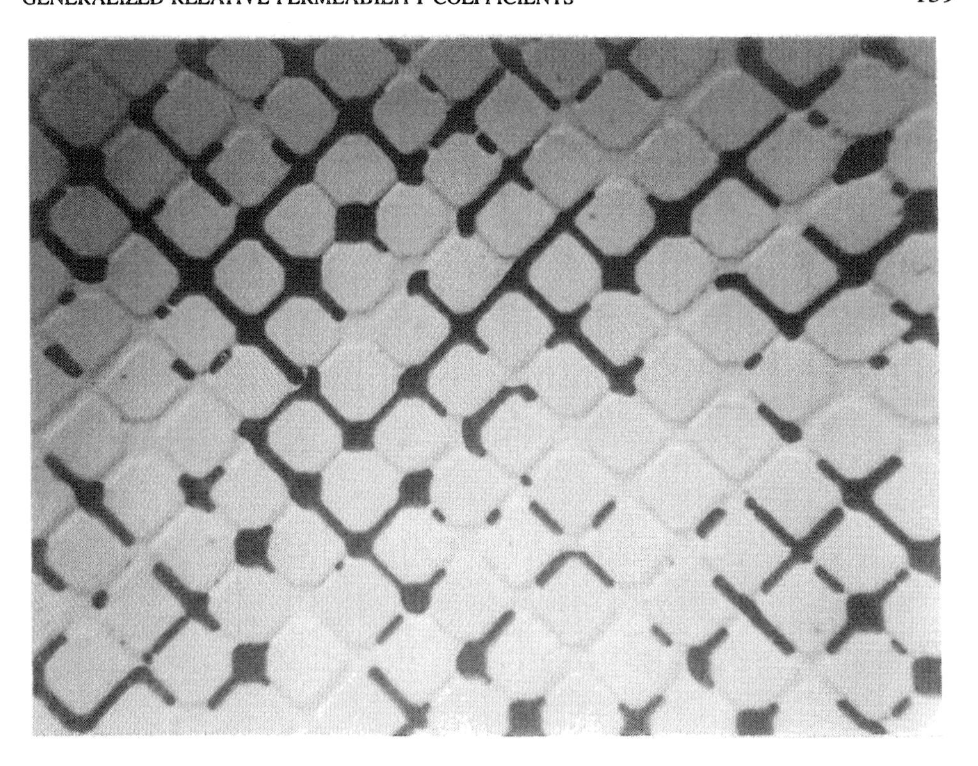

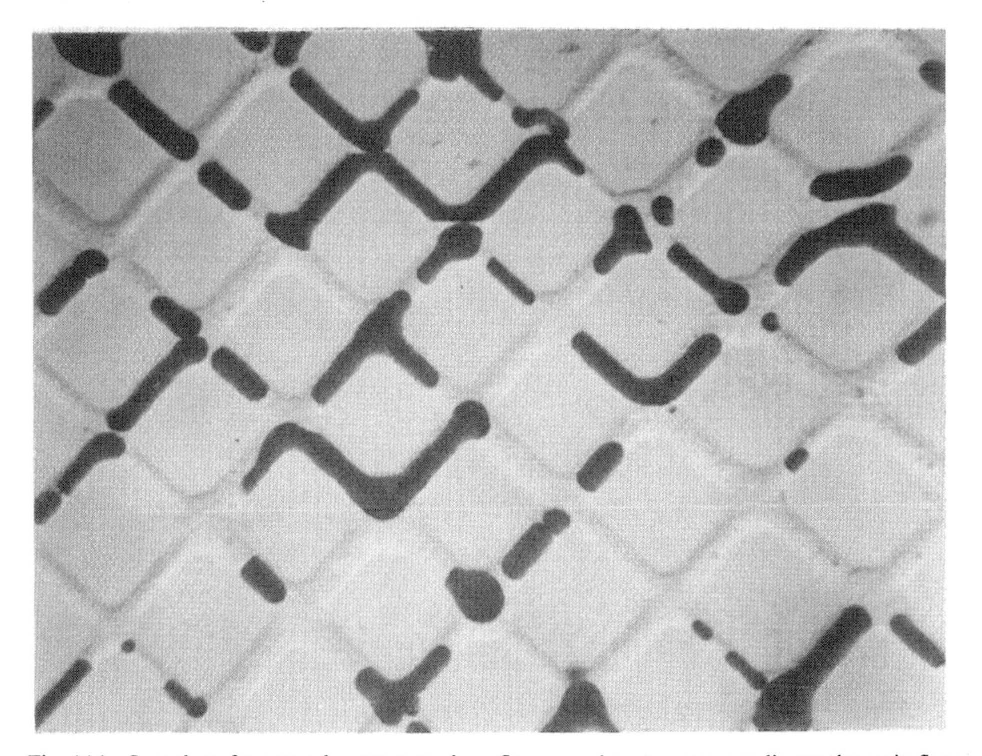

Fig. 1(a). Snapshots from steady-state two-phase flow experiments corresponding to the main flow regimes (a) Large ganglion dynamics (LGD). (1b) Small ganglion dynamics (SGD).

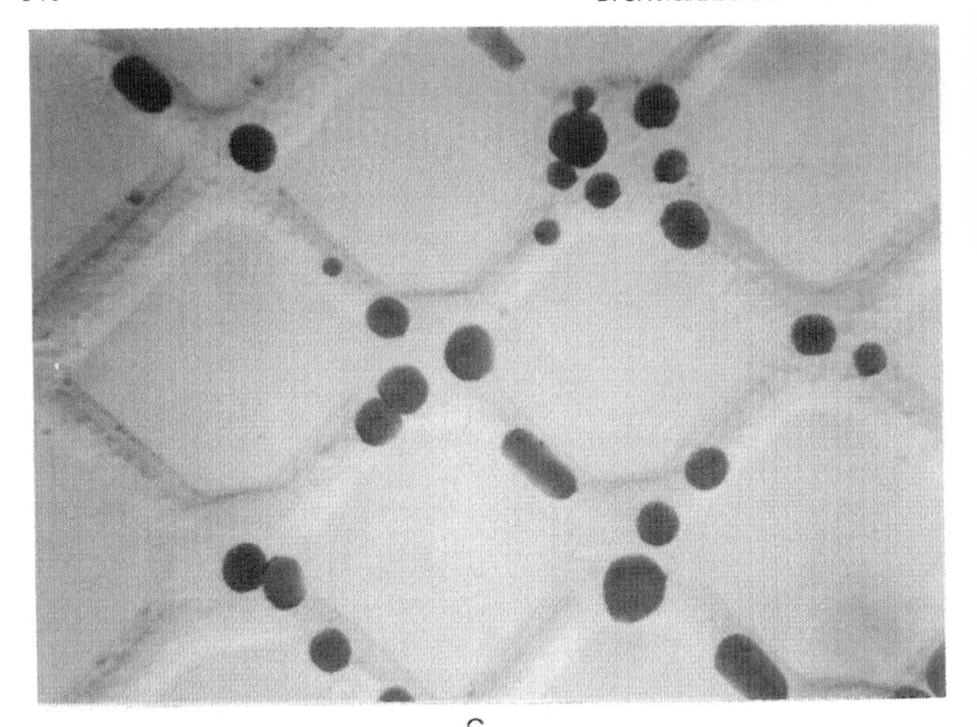

c

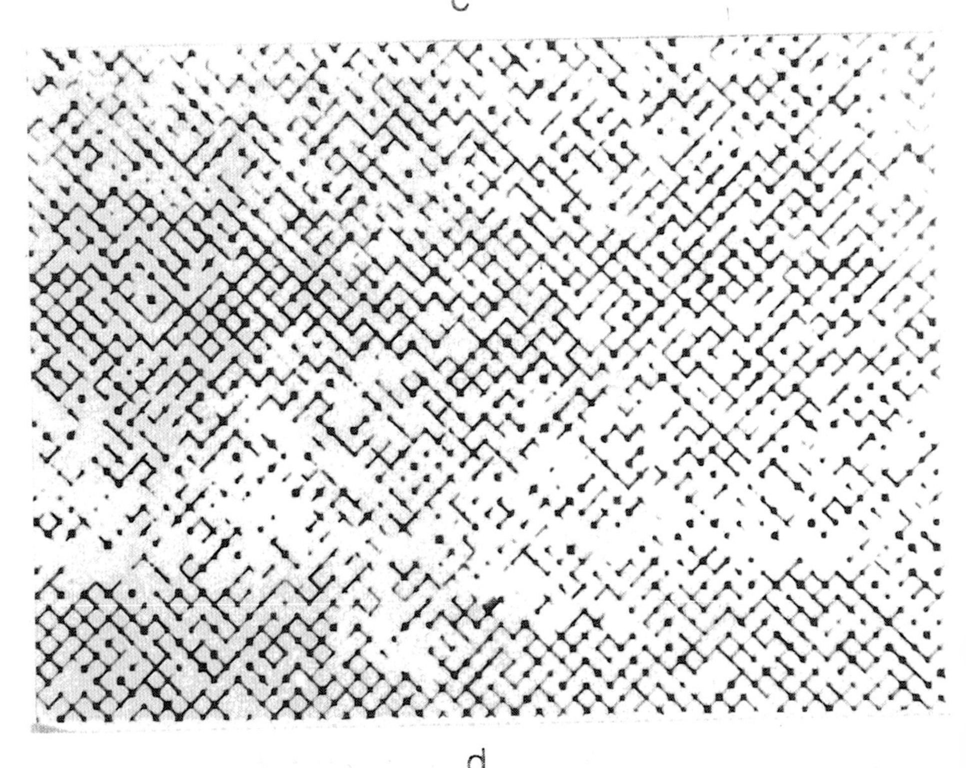

d

Fig. 1(c). Drop traffic flow (DTF) detail. (1d) Connected pathway flow (CPF). The snapshot shows the coexistence of connected pathway flow with small ganglion dynamics.

pathways, especially near the fringes of the oil pathways. The conventional relative permeabilities were found to correlate with the flow regimes strongly. The relative permeability to oil, k_{r2}, is minimal in the domain of LGD, and increases strongly as the flow mechanism changes from LGD to SGD to DTF to CPF. The relative permeability to water, k_{r1}, is minimal in the domain of SGD, and increases weekly as the flow mechanism changes from SGD to LGD, whereas it increases strongly as the mechanism changes from SGD to DTF to CPF.

To express the relative magnitude of the coupling effects, we define two new quantities, the *coupling indices*, as follows, $\chi_i \equiv 1-(k_{rii}/k_{ri})$. The experimentally determined values of the coupling coefficients were found to be correlated strongly with the flow regimes.

In steady-state two-phase flow the macroscopic pressure gradients in the two fluids are virtually equal, provided that the flow is fully developed (free from end-effects), because then the macroscopic gradient of the capillary pressure is nil. This is not only a theoretical expectation, based on the flow mechanisms, but also the result of careful experimental measurements (Bensten and Manai, 1993). In this case the coupling coefficients can be embedded in the conventional relative permeability coefficients, through the relations $[k^{\circ}_{ri} = k^{\circ}_{rii} + (\mu_i/\mu_j)k^{\circ}_{rij}; i, j = 1, 2; i \neq j]$. This can conceal the fact (and has done so many times in the past) that the coupling coefficients are important and that in many cases of practical interest they make the main contribution. The present work suggests a synthesis in the sense that one can use the conventional relative permeability coefficients to describe the macroscopic behaviour without losing sight of the fact that known and significant portions of their magnitudes (equal to the aforementioned coupling indices, χ_i) are due to coupling effects. This synthesis applies strictly to steady-state co-current fully-developed flow; it does not apply to situations involving significant macroscopic gradients of the saturation and the capillary pressure, such as displacement fronts, etc.

3. Conventional and Generalized Relative Permeability Coefficient: A Review

The problem of the two-phase flow in porous media is usually treated with Darcy's extended law. For *horizontal, one-dimensional, immiscible, two-phase* flow in *homogeneous* and *isotropic* porous media, these equations take the form

$$v_1 = -\frac{kk_{r1}}{\mu_1}\frac{\Delta P_1}{L}, \tag{1a}$$

$$v_2 = -\frac{kk_{r2}}{\mu_2}\frac{\Delta P_2}{L}, \tag{1b}$$

where k_{r1} and k_{r2} are the conventional relative permeabilities for fluids 1 and 2, respectively, and k is the absolute permeability of the porous medium. In these equations, the interactions between the fluids are neglected and, therefore, the conventional relative permeabilities are thought (not rigorously) to represent the drag due to the flow of each particular fluid over the solid surface. In effect, it is assumed (arbitrarily) that the solid surface and each one of the fluids form a new porous matrix through which the other fluid can flow. This implies that all fluid–fluid *interfaces* remain *static* in steady-state flow, which is *not true*.

Once we accept that the fluid-fluid interfaces are not static, the interactions of the fluids become significant and the drag due to momentum transfer across the fluid-fluid interfaces should be taken into account. This gives rise to the viscous coupling effects. It has been shown theoretically that the magnitude of viscous coupling depends on the rheology of the oil–water interface, and it diminishes when the interfacial shear viscosity becomes significant (Ehrlich, 1993). The governing equations for *coupled immiscible two-phase* flow in *homogeneous* and *isotropic* porous media have been suggested by many authors (Raats and Klute, 1968; Rose, 1972; Shanchez-Palencia, 1974; Marle, 1982; de Gennes, 1983; de la Cruz and Spanos, 1983; Whitaker, 1986; Spanos *et al.*, 1986; Kalaydjian, 1987; Auriault, 1987; Auriault *et al.*, 1989) and their integrated form can be written as

$$v_1 = -\frac{kk_{r11}}{\mu_1}\frac{\Delta P_1}{L} - \frac{kk_{r12}}{\mu_2}\frac{\Delta P_2}{L}, \tag{2a}$$

$$v_2 = -\frac{kk_{r21}}{\mu_1}\frac{\Delta P_1}{L} - \frac{kk_{r22}}{\mu_2}\frac{\Delta P_2}{L}, \tag{2b}$$

where k_{r11} and k_{r22} are the generalized relative permeabilities for phases 1 and 2, respectively, and k_{r12} and k_{r21} are the viscous coupling coefficients.

The relative importance of the viscous coupling coefficients has been debated widely in the literature. The coupling coefficients are often assumed to be significant only when $\mu_1 = \mu_2$ and the saturation of one fluid is relatively low, but there is experimental evidence that the viscous coupling coefficients are also important when the fluid viscosities differ substantially (Kalaydjian and Legait, 1987; Goode and Ramakrishnan, 1993; also the present work). The coupling effects were reported to be insignificant in Yadav *et al.* (1987), where the conventional relative permeabilities of both phases remained unchanged during a drainage process when the opposite phase was solidified in situ. However, recent experimental studies (Kalaydjian *et al.*, 1989; Bourbiaux and Kalaydjian, 1988, 1990; Bentsen and Manai, 1993) have shown that the viscous coupling coefficients are indeed important and comparable to the diagonal coefficients.

Indirect but strong evidence that viscous coupling is important is that the conventional relative permeabilities depend on almost all the pertinent flow parameters. Besides the saturation and saturation history of the fluids (Johnson *et al.*, 1959; Naar *et al.*, 1962; Jerault and Salter, 1990) they also depend on the capillary number, Ca (Leverett, 1939; Sandberg *et al.*, 1958; Taber, 1969; Lefebvre du Prey, 1973; Heavyside *et al.* 1982; Amaefule and Handy, 1982; Fulcher *et al.*, 1985), the viscosity ratio κ (Yuster, 1951; Odeh, 1959; Lefebvre du Prey, 1973), and the wettability characteristics, expressed here by the equilibrium contact angle θ_e (Owens and Archer, 1971; McCaffrey and Bennion, 1974; Morrow and McCaffery, 1978). Another important parameter is the coalescence factor, Co (defined as the mean coalescence probability given a collision between two interfaces) (Constantinides and Payatakes, 1990). Avraam and Payatakes (1995) reported, for the first time, a strong correlation between the conventional relative permeabilities and the corresponding steady-state two-phase flow mechanisms.

The following conclusions can be drawn. If one insists on using the conventional fractional flow theory, then one should modify the approach by including the dependence of the relative permeabilities on all the pertinent parameters, i.e., for a given porous medium,

$$k_{ri} = k_{ri}(S_i; Ca, \kappa, \cos\,\theta_e, Co, \text{saturation history}, ...). \tag{3}$$

The fact that k_1 depends *strongly* on Ca, which is proportional to v_1, implies that the relation between v_1 and ΔP_1 is strongly *nonlinear*, despite the fact that inertial effects are negligible under typical flooding conditions. The same holds true for v_2 and ΔP_2. If, on the other hand, one intends to use the generalized fractional flow theory, one needs a reliable method to determine the non-diagonal terms, taking into account that the generalized relative permeability coefficients also depend on the parameters appearing in Equation (3),

$$k_{rij} = k_{rij}(S_i; Ca, \kappa, \cos\,\theta_e, Co, \text{saturation history}, ...). \tag{4}$$

Conventional relative permeabilities are measured in the laboratory with two basic methods: *unsteady-state flow* experiments (i.e. *immiscible displacement* experiments), and steady-state flow experiments.

The techniques used to estimate the conventional relative permeabilities from immiscible displacement data introduce many uncertainties. All techniques originate from the Buckley–Leverett theory (Buckley and Leverett, 1941; Rapoport and Leas, 1952). Those techniques, which are called explicit (Welge, 1952; Johnson *et al.*, 1959; Jones and Roszelle, 1978), calculate the point values of the conventional relative permeabilities directly from the measured data. However, they are subjected to significant errors (Tao and Watson, 1984). Implicit techniques may be used, instead. According to these techniques, a functional representation for the conventional relative permeabilities is chosen so that the effluent data evaluated by the mathematical model, with or without the capillary pressure, match the experimental data in a statistical sense. The functional representations may be simple equations with two parameters (Archer and Wong, 1973; Sigmund and McCaffery, 1979; Batycky *et al.*, 1981), but representations with cubic splines or B-splines are more flexible and reliable (Kerig and Watson, 1986; Watson *et al.*, 1988) and reduce significantly the bias error. The methods of estimating the unknown parametres can be simple trial-and-error adjustment (Archer and Wong, 1973), or general nonlinear regression techniques. The latter permit the analysis of the variance error of the estimates and make possible the physical interpretation of them. Despite the sophistication of these techniques, it is highly debatable whether the concept of relative permeabilities can be applied rigorously in a narrow front involving steep and fuzzy gradients of saturation and pressure.

Steady-state flow experiments can be used to estimate conventional relative permeabilities much more reliably. The immiscible fluids flow simultaneously until saturation and pressure equilibrium is attained. This is usually time-consuming, which is a considerable disadvantage. Relative permeability data at various values of saturation are taken in a stepwise manner by adjusting the flowrate ratio appropriately, and this is also time-consuming. On the other hand, data produced by this method are more reliable because the saturations, fluid flow rates, and pressure gradients are all directly measured and correlated using Darcy's law. A suitable modification of the experimental method should also take into account the flow mechanisms and the pertinent flow parameters, as is suggested in Avraam and Payatakes (1995).

When one intends to determine generalized relative permeability coefficients from steady-state flow experiments, one set of experiments does not provide sufficient information for the unique determination of the four coefficients (Whitaker, 1986). A second set of experiments is necessary. Rose (1988) suggested that two sets of steady-state experiments, one without gravitational effects and another with gravitational effects, would provide sufficient data. However, this procedure is subject to great errors (Rose, 1989). Bourbiaux and Kalaydjian (1988) performed two sets of experiments, one with co-current two-phase flow with no capillary effects (i.e., with high flow rates) and another with countercurrent two-phase flow and null total flow, as it was first proposed by Lelievre (1966). They calculated the generalized coefficients by assuming that the viscous coupling coefficients were the same in the two types of experiments. A similar experimental procedure was followed by Kalaydjian (1990) in a square cross-section capillary tube and it was suggested for application in real porous media. However, this procedure is open to criticism because the two different injection modes (co-current and counter-current) produce different flow configurations, and the generalized permeability coefficients should be affected accordingly. In all these studies, the two coupling coefficients were taken equal to each other ($k_{r12}/\mu_2 = k_{r21}/\mu_1$), based on the assumption that the Osanger–Casimir reciprocity equation applies. Thus, only three coefficients had to be determined each time. This assumption needs to be discussed. In flow regimes such as 'steady-state' ganglion dynamics (Avraam and Payatakes, 1995), in which the motion of oil involves continual occurrence of *catastrophic pore-level flow events* (ganglion breakup, coalescence, stranding, mobilization), *microscopic reversibility does not exist*, and the Onsager–Casimir reciprocal relation should not be expected to hold. In Kalaydjian's experiments using square cross-section capillary tubes, catastrophic events were absent, and therefore his use of the reciprocity relation was justified. One, however, should not make this assumption in more complex situations, such as those already mentioned. Indeed, Bentsen and Manai (1993) showed that the two off-diagonal viscous coupling coefficients are importants and not necessarily equal. The present work also shows that the off diagonal coefficients are very important and very unequal. A final comment is that to date only the dependence of the generalized coefficients on the fluid saturation has been studied. An extension including the rest of the flow parameters is warranted.

In this work, a proper combination of co-current steady-state two-phase flow experiments with a nonlinear parameter estimation method is suggested as a reliable and convenient means for the determination of all the necessary coefficients. The method has the advantage that one set of experiments (which are the same in nature with the flow process under consideration) is sufficient, and that the estimated coefficients can be endowed with a consistent physical interpretation. This gives the possibility, for the first time, to correlate all coefficients with the pertinent flow parameters, as well as with the underlying flow mechanisms.

4. Theory

4.1. PARAMETER ESTIMATION METHOD

In the context of the parameter estimation method, the dependent variables of the model under consideration are represented with a functional form,

$$\mathbf{y} = \mathbf{f}(\mathbf{x}; \boldsymbol{\theta}), \tag{5}$$

where $\mathbf{y}$ and $\mathbf{x}$ are the vectors of the dependent and the independent variables of the model, respectively, and $\boldsymbol{\theta}$ is the vector of the unknown parameters. A suitable experimental method provides the necessary measurements for both dependent and independent variables, $\mathbf{y}$ and $\mathbf{x}$. The measured values that correspond to the μth experiment and the αth equation of the model are denoted by $y_{\mu\alpha}$ and $x_{\mu\alpha}$. The unknown parameters $\{\theta_i, i = 1, \ldots, N\}$, where N is the total number of parameters in the model, are chosen so as to minimize the sum of the squares of the residuals,

$$e_{\mu\alpha} = y_{\mu\alpha} - f_{\mu\alpha}(\mathbf{x}_\mu; \boldsymbol{\theta}). \tag{6}$$

The sum of the squared residuals gives the objective function, which has the form

$$\Phi(\boldsymbol{\theta}) = [\mathbf{y} - \mathbf{f}(\mathbf{x}; \boldsymbol{\theta})]^T \mathbf{W} [\mathbf{y} - \mathbf{f}(\mathbf{x}; \boldsymbol{\theta})] = \mathbf{e}^T \mathbf{W} \mathbf{e}, \tag{7}$$

where $\mathbf{W}$ is a weighing matrix and $\mathbf{e}$ is the vector of the residuals, $e_{\mu\alpha}$.

The value of the parameter vector that minimizes the objective function, $\boldsymbol{\theta} = \boldsymbol{\theta}^*$, is called the estimator of the parameters. According to the selection of the weighing matrix, $\mathbf{W}$, the estimates of the parameters, $\boldsymbol{\theta}_i$, acquire certain desirable properties. If $\hat{\boldsymbol{\theta}}$ is the true (but unknown) value of the parameter vector, and we denote with $\mathbf{V}$ the covariance matrix of the true errors, $\varepsilon_{\mu\alpha} = y_{\mu\alpha} - f_{\mu\alpha}(\mathbf{x}_\mu; \hat{\boldsymbol{\theta}})$, for all experiments and equations of the model, we can choose $\mathbf{W} = \mathbf{V}^{-1}$. Then the maximum-likelihood estimator is obtained, which maximizes the corresponding likelihood (Bard, 1974). Maximum likelihood estimators have important properties; they are consistent (i.e., asymptotically unbiased), and asymptotically efficient (i.e., as the number of the measurements increases beyond bound, the attainable covariance matrix of the estimator tends to a lower bound, given by the Rao–Cramer theorem).

Usually, the covariance matrix of the errors is not known, and some additional assumptions have to be made concerning the nature of the errors. If the errors are assumed to be independent and to have zero mean, the covariance matrix reduces to a diagonal matrix, whose elements are the variances of the errors. Then, the objective function takes the simpler form:

$$\Phi(\boldsymbol{\theta}) = \sum_{\alpha=1}^{a} \sum_{\mu=1}^{n} \frac{1}{\sigma_{\mu\alpha}^2} [y_{\mu\alpha} - f(\mathbf{x}_{\mu\alpha}; \boldsymbol{\theta})]^2 \tag{8}$$

where, $\sigma_{\mu\alpha}^2$ is the variance of the error in the μth experiment and in the αth equation, a is the number of the model equations, and n is the number of the available experimental measurements. A more general form could be obtained, if the covariance matrix of the errors were nondiagonal, but the simple form of the objective function in Equation (8) is sufficient for the estimation of the parameters.

Finally, if the variances of the diagonal covariance matrix are unknown, the stagewise maximization method can be used (Bard, 1974). According to this method, the unknown variances and the parameters are estimated simultaneously so as to minimize the objective function. The method can also be used when the covariance matrix of the errors has the general nondiagonal form.

The two models of the macroscopic equations of two-phase flow in porous media, i.e. the conventional one given by Equations (1a), (1b), and the generalized one given by Equations (2a), (2b), can be put directly in the form of Equation (5). We choose as independent variables the fluid pressure drops, ΔP_1 and ΔP_2, and the water saturation, S_1, and as dependent variables the superficial fluid velocities, v_1 and v_2, respectively. The functional representations of the conventional relative permeabilities and of the generalized relative permeability coefficients, written explicitly, should satisfy Equations (3) and (4). If we assume that the unknown covariance matrix of the errors is diagonal, the maximum likelihood estimator minimizes the following objective functions for the two models, respectively:

$$\Phi^{(1)}(\boldsymbol{\theta}) = \sum_{\mu=1}^{n} \left\{ \left[v_{1\mu} + \frac{k k_{r1}(S_{1\mu}; \boldsymbol{\theta})}{\mu_1} \Delta P_{1\mu} \right]^2 \right\} + \\ + \sum_{\mu=1}^{n} \left\{ \left[v_{2\mu} + \frac{k k_{r2}(S_{1\mu}; \boldsymbol{\theta})}{\mu_2} \Delta P_{2\mu} \right]^2 \right\} \tag{9a}$$

and

$$\Phi^{(2)}(\boldsymbol{\theta}) = \sum_{\mu=1}^{n} \left\{ \left[v_{1\mu} + \frac{kk_{r11}(S_{1\mu};\boldsymbol{\theta})}{\mu_1} \Delta P_{1\mu} + \frac{kk_{r12}(S_{1\mu};\boldsymbol{\theta})}{\mu_2} \Delta P_{2\mu} \right]^2 \right\} +$$

$$+ \sum_{\mu=1}^{n} \left\{ \left[v_{2\mu} + \frac{kk_{r21}(S_{1\mu};\boldsymbol{\theta})}{\mu_1} \Delta P_{1\mu} + \frac{kk_{r22}(S_{1\mu};\boldsymbol{\theta})}{\mu_2} \Delta P_{2\mu} \right]^2 \right\}. \quad (9b)$$

In Equations (9a) and (9b) only the dependence of the coefficients k_{ri} and k_{rij} on S_1 (and of course on $\boldsymbol{\theta}$) is written explicitly. This is done to avoid cumbersome notation. The dependence of the coefficients on the rest of the flow parameters, Ca, κ, θ_e, Co, saturation history, porous medium, etc., which is not explicitly written, has been considered separately. To this end, we made a parametric experimental study by changing systematically the values of Ca and κ, while keeping the rest of the parameters constant. Then, the coefficients were estimated independently from each different set of experimental measurements obtained with a different set of Ca and κ values. The number of the experimental data available for each set of the flow parameters was either 5 (in most cases) or 4, as taken from Avraam and Payatakes 1995).

4.2. FUNCTIONAL REPRESENTATION OF THE TRANSPORT COEFFICIENTS

The problem of estimating the generalized relative permeability coefficients in Equations (9a) and (9b) is put in the form of a parameter estimation problem, when a suitable functional form is selected to represent the coefficients. By making the proper selection of the functional representation, one can minimize the total estimation error, which is composed of the bias error and the variance error. The former is due to the inability of the functional form to represent the true (albeit unknown) form of the coefficients, whereas the latter is due to the statistical uncertainty of the estimates caused by the inexact nature of the experimental measurements. Usually, a functional form with a great number of parameters decreases the bias error, since it is capable or representing a greater number of functions. On the other hand, it increases the variance error. A proper selection should minimize the total estimation error, which however is generally unknown.

Polynomial splines, or B-splines, are highly flexible functional forms which reduce drastically the bias error (Kerig and Watson, 1986). On the other hand, the variance error can be kept small through proper selection of the number of unknown parameters (Watson *et al.*, 1988). Here, we use cubic B-splines to represent the conventional relative permeabilities and the generalized relative permeability coefficients, and a suitable method in order to select the proper number of unknown parameters:

$$k_{ri} = \sum_{k=1}^{N} C_k^i \phi_k(S_1) \quad (10a)$$

and

$$k_{rij} = \sum_{k=1}^{N} C_k^{ij} \phi_k(S_1), \tag{10b}$$

where C_k^i and C_k^{ij} are the unknown parameters for k_{ri} and k_{rij}, respectively, $\{\phi_k; k = 1, 2, \ldots, N\}$, are basic cubic B-splines, and N is the number of the unknown parameters. For the sake of simplicity, we keep the same number of parameters for all coefficients, though this is not necessary.

By substituting Equations (10a) and (10b) in the objective functions (9a) and (9b), the parameters C_k^i and C_k^{ij} form the respective vectors $\boldsymbol{\theta}$, and the objective functions are to be minimized with respect to $\boldsymbol{\theta}$.

4.3. OPTIMIZATION METHOD

We used the standard Gauss–Marquardt method to minimize the objective functions (9a) and (9b). The method was found to be fast and stable. With four unknown parameters the method usually needed less than 100 iterative steps to converge.

Although it is always possible to find an (at least local) unconstrained minimum of the objective functions, the physical meaning of the coefficients requires that these coefficients should be positive. For this reason, the following linear inequality constraints are imposed:

$$k_{ri}(S_1, \boldsymbol{\theta}) \geqslant 0, \quad i = 1, 2 \tag{11a}$$

and

$$k_{rij}(S_1, \boldsymbol{\theta}) \geqslant 0, \quad i, j = 1, 2. \tag{11b}$$

For the problem of constrained minimization we used a projection method suitable for linear constraints. The projection method follows the Gauss–Marquardt minimization step whenever the coefficients are well inside the feasible region, whereas it follows the boundaries of the feasible region when a coefficient attempts to 'fall off'. In the latter case, a quadratic programming (QP) method is implemented, which performs constrained minimization of the local quadratic approximation of the objective function. The method uses the Kuhn–Tucker conditions (see Bard, 1974) to find the solution of the QP problem along the active constraints (i.e. constraints for which the equality holds locally). This also ensures the constrained minimization of the objective function, when the minimum is found to be on the boundaries of the feasible region. We found that the latter is not always the case, but in some cases the minimum is found inside the feasible region.

4.4. SELECTION OF THE APPROPRIATE NUMBER OF PARAMETERS

As the number of the unknown parameters increases the bias error (although unknown) decreases. On the other hand, the total number of parameters is limited by the number of the available experimental measurements. The general rule is that the number of the experimental measurements must exceed the maximum number of unknown parameters per equation of the model, when the (diagonal) covariance matrix of the errors is unknown (Bard, 1970):

$$n > \max(l_\alpha). \tag{12}$$

In the present work and for a given set of conditions (fixed Ca, κ, etc.), we had $n = 5$ experimental measurements in the cases of $Ca = 10^{-7}$, 10^{-6}, and $n = 4$ in the case of $Ca = 5 \times 10^{-6}$. For the sake of simplicity, all relative permeability coefficients were represented with four B-splines ($N = 4$). Such representations are simple and easy to handle. Thus, for the estimation of the conventional relative permeability coefficients, the number of the unknown parameters per equation was $l_\alpha = 4$, that is, one less than the number of the experimental measurements. For the case $Ca = 5 \times 10^{-6}$ we produced one additional artificial experimental measurement using the Aitken interpolation method. Such artificial data are known as interpolation data. For the estimation of the generalized coefficients, the number of the unknown parameters per equation was $l_\alpha = 8$. Using the Aitken method we produced four or five interpolation data points for $Ca = 10^{-7}$, 10^{-6} and $Ca = 5 \times 10^{-6}$ respectively, and the number of unknown parameters remained always one less than the number of the experimental measurements.

Figure 2a shows the convergence of the estimated conventional relative permeability coefficients, k_{r1} and k_{r2}, when the number of parameters increases approaching that of the experimental measurements ($n = 5$). The estimated values improve from $N = 1$ to $N = 4$ and the objective function decreases as in Watson *et al.* (1988). When the number of parameters and experimental measurements are equal ($N = n = 5$) the parameter estimation is meaningless, since an exact solution of the model equations is possible. In that case the likelihood diverges.

The results of the estimation procedure, when interpolation data are produced for $n = 10, 20, 40$ and $N = 4$, are shown in Figure 2b. Figure 2c shows the results of the estimation procedure keeping the number of the experimental data the same as in Figure 2b but using $N = n - 1$. We observe that no appreciable improvement in the estimated values is achieved, which means that interpolation data are not necessary. The same holds for the generalized relative permeability coefficients; here, $n = 9$ experimental and interpolation data were sufficient for the estimation (see below, Figure 6).

5. Results and Discussion

5.1. EXPERIMENTAL APPARATUS AND PROCEDURE

A detailed description of the experimental apparatus and procedure is given in Avraam and Payatakes (1995). For the sake of self-sufficiency of the present work a brief description is given below.

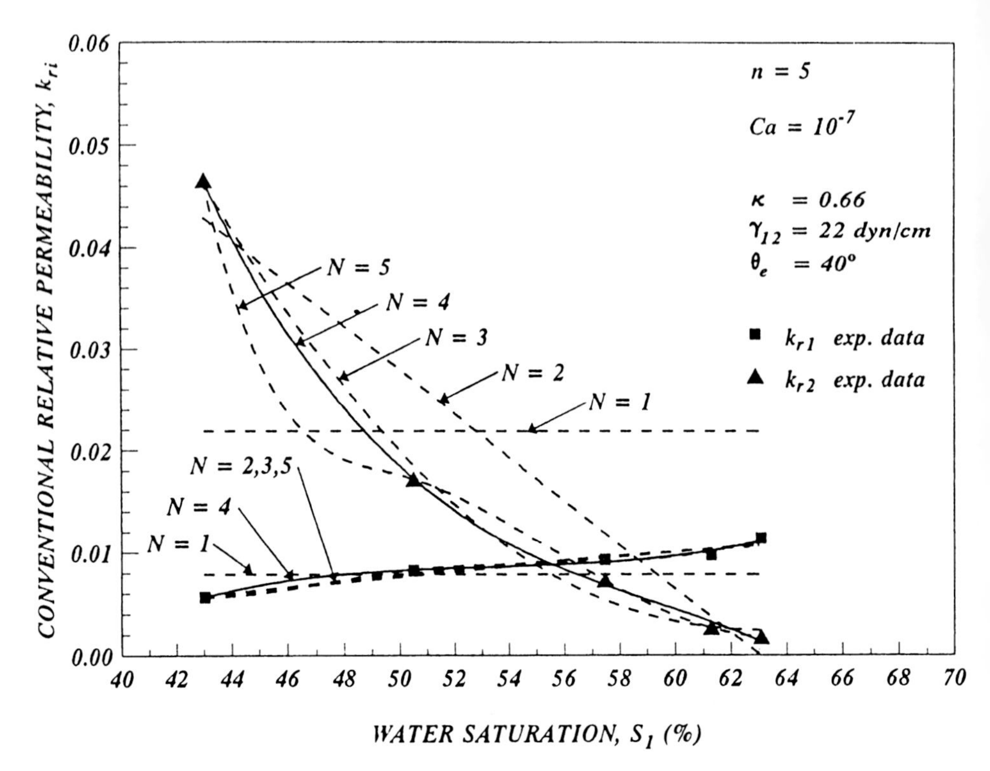

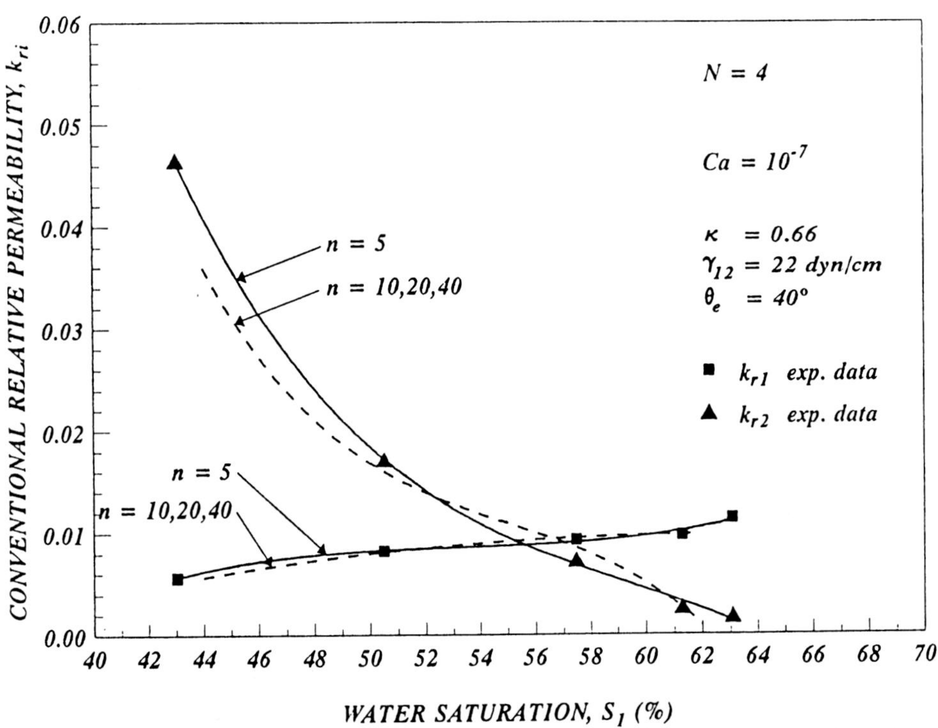

Fig. 2. Estimated curves of the conventional relative permeability coefficients, for a given set of flow conditions ($Ca = 10^{-7}$, $\kappa = 0.66$). (a) Dependence of the estimated curves on the number of parameters used in the B-spline representation of the coefficients. (b) Dependence of the estimated curves on the number of experimental data when interpolation data are added to the actual experimental measurements. The functional representation of the coefficients contains four parameters.

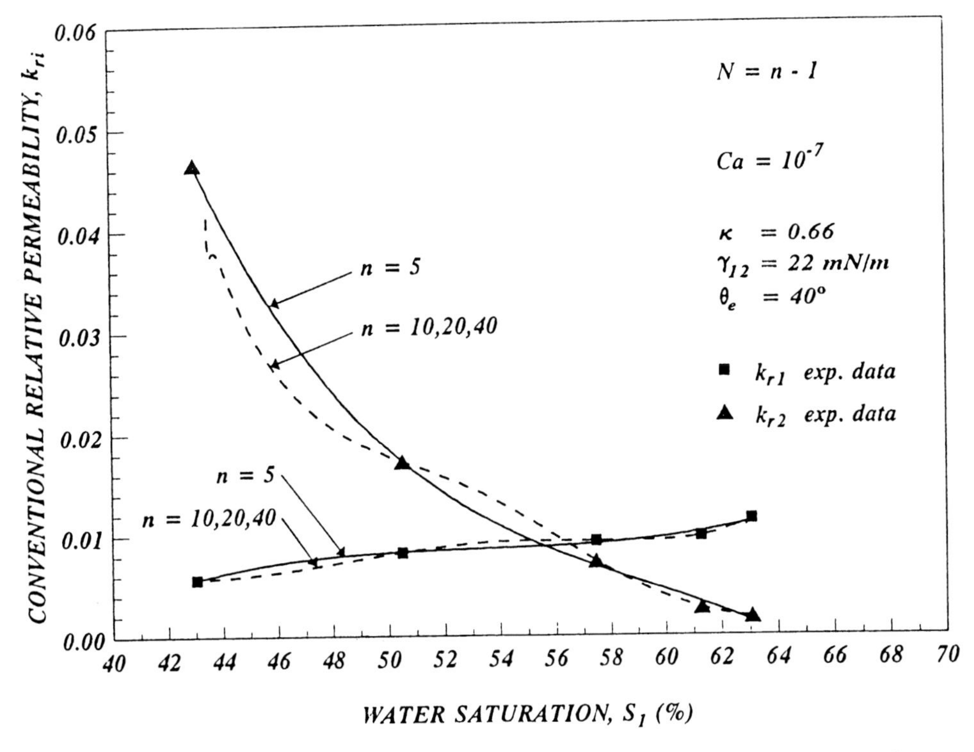

Fig. 2. (c) Same with b, but the functional representation of the coefficients contains one parameter less than the number of experimental data.

The model porous medium is a network of the chamber-and-throat type, etched in glass, and has characteristics that make it a relatively simple representation of a reservoir sandstone. It comprises 11 300 chambers and 22 600 throats. The mean values of the chamber and throat diameters are 560 μm and 112 μm, respectively, and the standard deviations of the populations are ca. 1/4 of the corresponding mean values. The maximum pore depth is nearly uniform and equal to ca. 140 μm. The node-to-node distance of the network is $l = 1221$ μm. At the beginning of each experiment the network is completely filled with the nonwetting liquid. Then, the two fluids are injected simultaneously with constant flowrates, using two syringe pumps, until steady-state is attained.

At steady-state measurements are made over a *central* region that covers ca. 1/4 of the entire network, to reduce the influence of end-effects. The pressure drop in each liquid is measured separately. This is achieved by placing suitable semipermeable membranes at the ends of the pressure taps for each of the two liquids. The pressure differences are monitored and recorded with a sensitive pressure transducer system, and the pressure signals are time-averaged to produce smooth, steady signals. The flow mechanisms are observed and videorecorded. The videorecorded signal is then digitized and processed with an image analyzer to obtain the saturation values of the two fluids as functions of position and time. These data, in turn, are space-averaged and time-averaged to obtain the mean values of S_1 and S_2 corresponding to the conditions of the experiment. The end-effects do not significantly affect the details of the flow mechanisms in the region of measurements. This was verified by partitioning the region of measurements in six segments and by making statistical analysis of the time-averaged observations concerning the 'oil' (number of ganglia per unit area, ganglion size distribution) to check for any significant differences among the segments. None were found.

The data obtained in this way were processed to obtain the results presented below. It should be noted that the capillary number is calculated from $Ca = \mu_1 q_1/\gamma_{12} wl$, where q_1 is the flow rate of the 'water' and $w (= 70\,\text{mm})$ the width of the network.

5.2. FLOW REGIMES AND GENERALIZED RELATIVE PERMEABILITY COEFFICIENTS

Photographs of the main flow regimes of steady-state co-current two-phase flow are shown in Figures 1a–d. Figure 1a shows LGD (large ganglion dynamics), Figure 1b shows SGD (small ganglion dynamics), Figure 1c shows DTF (drop traffic flow), and Figure 1d shows CPF and SGD (connected pathway flow and SGD).

Figures 3, 4 and 5 present all the estimated values of the conventional (dashed lines) and generalized (solid lines) relative permeability coefficients as functions of the water saturation, S_1. Each figure corresponds to a constant value of the viscosity ratio, $\kappa = 0.66, 1.45, 3.35$, respectively (a fixed pair of immiscible fluids), whereas the capillary number, Ca, increased going from (a) to (c) and from 10^{-7} to 10^{-6} to 5×10^{-6}. The other flow parameters (θ_e, Co, Bo, etc.) are kept constant. The flow regimes that correspond to the specific ranges of S_1 and Ca are indicated (see also Figure 7).

The conventional and the generalized coefficients depend on the flow conditions and the corresponding flow mechanics. Summarizing, we can say that the conventional relative permeability coefficients, k_{r1} and k_{r2}, are increasing functions of the saturation of the respective fluid, and they increase as Ca and/or κ increase. The generalized relative coefficient k_{r12} increases with increasing S_1, whereas both k_{r21} and k_{r22} decrease with increasing S_1. The behavior of k_{r11} is more complex. In most cases studied (Figures 3a, 3b, 4a, 5a, 5b, 5c) k_{r11} increases as S_1 increases. However, in certain cases (Figures 3c, 4c) k_{r11} decreases as S_1 increases, whereas in others (Figure 4b) it displays mixed behavior. As Ca increases we can say, roughly, that all the generalized coefficients increase.

With regard to the flow of water, k_{r11} is the dominant term in the domain of CPF, where it provides the main contribution to the conventional relative permeability k_{r1}. An exception to this is observed in the domain of CPF and SGD, at large S_1 values, where k_{r12} becomes significant. This is an indication that the flow of the water is assisted by the motion of the numerous ganglia (and droplets) that move in

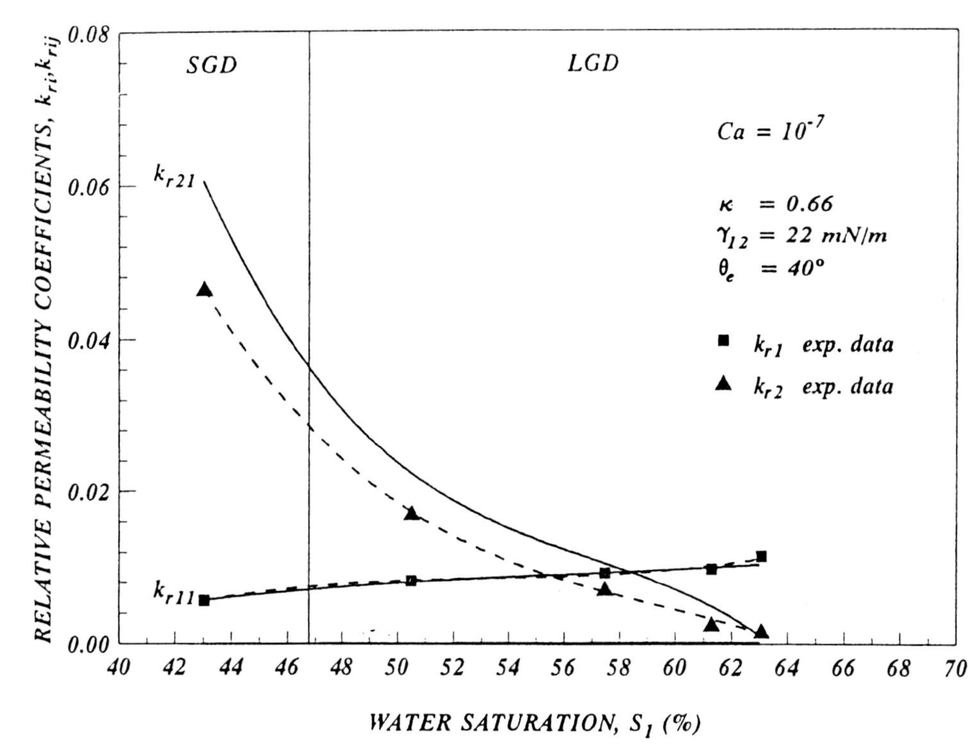
SGD
LGD
Ca = 10^-7
κ = 0.66
γ12 = 22 mN/m
θe = 40°
kr1 exp. data
kr2 exp. data
kr21
kr11
RELATIVE PERMEABILITY COEFFICIENTS, kri,krij
WATER SATURATION, S1 (%)
0.00
0.02
0.04
0.06
0.08
40
42
44
46
48
50
52
54
56
58
60
62
64
66
68
70

Fig. 3a.

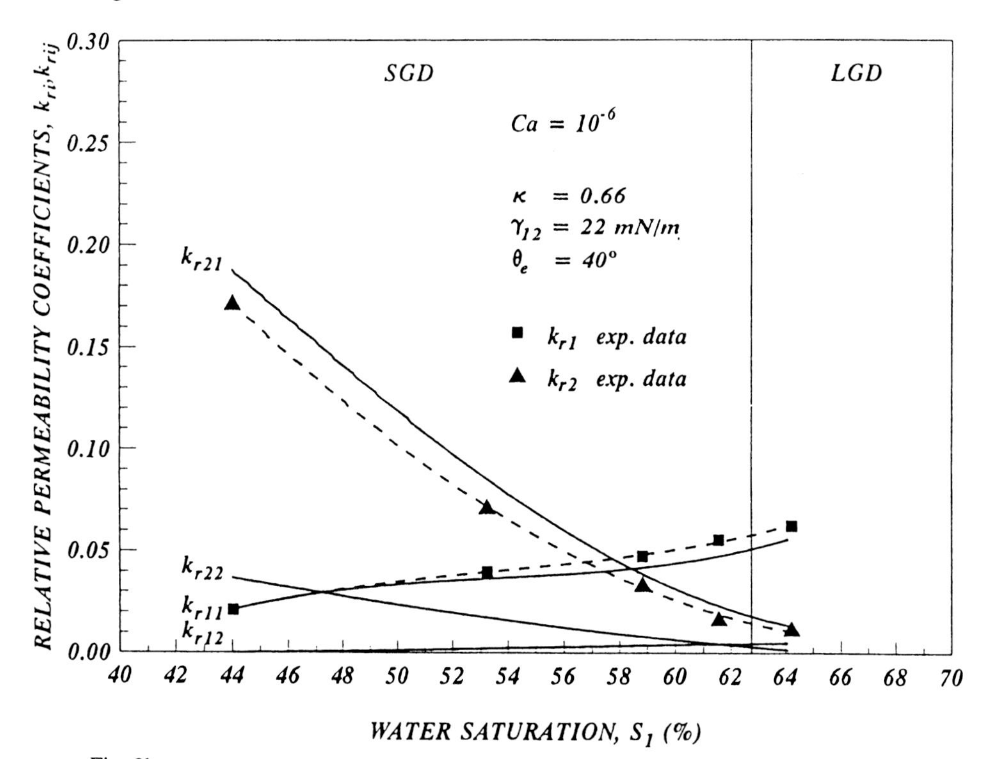
SGD
LGD
Ca = 10^-6
κ = 0.66
γ12 = 22 mN/m
θe = 40°
kr1 exp. data
kr2 exp. data
kr21
kr22
kr11
kr12
RELATIVE PERMEABILITY COEFFICIENTS, kri,krij
WATER SATURATION, S1 (%)
0.00
0.05
0.10
0.15
0.20
0.25
0.30
40
42
44
46
48
50
52
54
56
58
60
62
64
66
68
70

Fig. 3b.

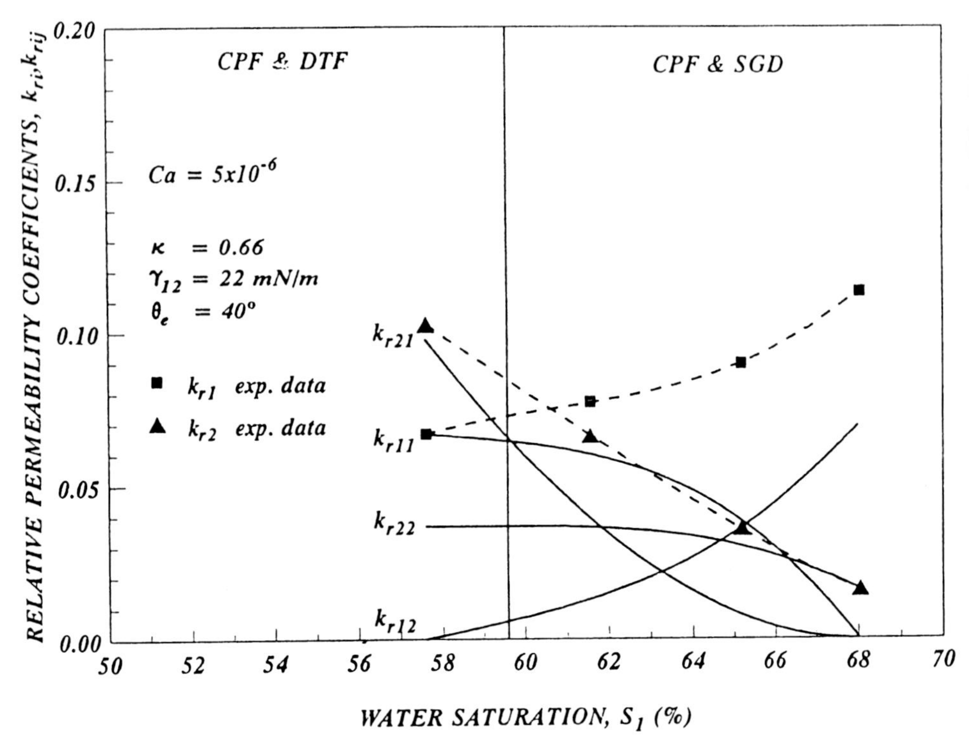

Fig. 3c.

Fig. 3. Conventional (dashed lines) and generalized (solid lines) relative permeability coefficients for $\kappa = 0.66$, with all the other physicochemical properties kept constant. (a) $Ca = 10^{-7}$, (b) $Ca = 10^{-6}$, (c) $Ca = 5 \times 10^{-6}$.

the water pathways, especially along the fringes of the connected oil pathways. In the domains of steady-state ganglion dynamics (LGD, SGD) the behavior is more complex. For $\kappa < 1$, the coefficient k_{r11} makes the dominant contribution, even though k_{r12} is discernible. For $\kappa > 1$ the coefficient k_{r11} becomes less significant, and k_{r12} becomes large, even dominant. This is strong evidence that the flow of the water is assisted considerably by the flow of the population of moving ganglia. This phenomenon is enhanced slightly as S_2 increases (S_1 decreases).

A significant result concerning the flow of the oil is that in the domains of LGD, SGD and DTF, where the oil is disconnected, k_{r21} is often nearly equal to the conventional relative permeability, k_{r2} (Figs. 3, 4, 5), whereas k_{r22} is small or nil. This shows the crucial importance of the coupling effects. In these cases the motion of the disconnected oil is caused mainly by the flowing water. The diagonal term k_{r22} becomes significant only at high oil saturation, S_2, especially in the domain of CPF and SGD, or CPF and DTF. In the latter case, the main contribution to the overall flow rate is that of the connected oil pathways, while the ganglia and/or droplets that move along the fringes of the oil pathways contribute to a smaller but still appreciable extent.

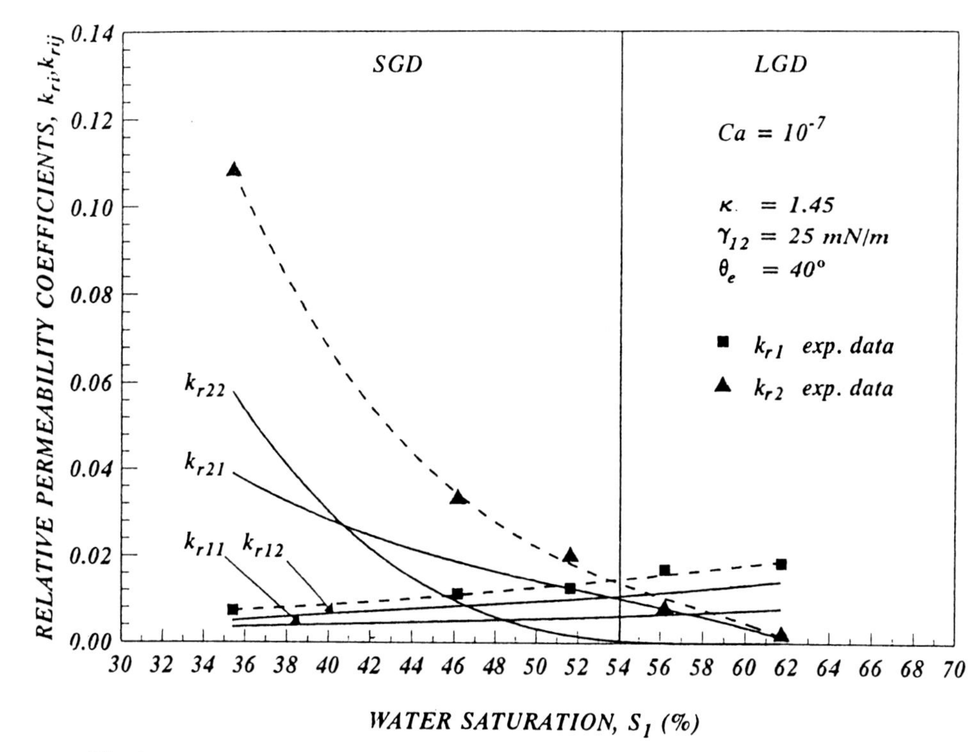

RELATIVE PERMEABILITY COEFFICIENTS, k_{ri}, k_{rij}
0.14
0.12
0.10
0.08
0.06
0.04
0.02
0.00
SGD
LGD
$Ca = 10^{-7}$
$\kappa = 1.45$
$\gamma_{12} = 25\ mN/m$
$\theta_e = 40°$
k_{r1} exp. data
k_{r2} exp. data
k_{r22}
k_{r21}
k_{r11}
k_{r12}
30 32 34 36 38 40 42 44 46 48 50 52 54 56 58 60 62 64 66 68 70
WATER SATURATION, S_1 (%)

Fig. 4a.

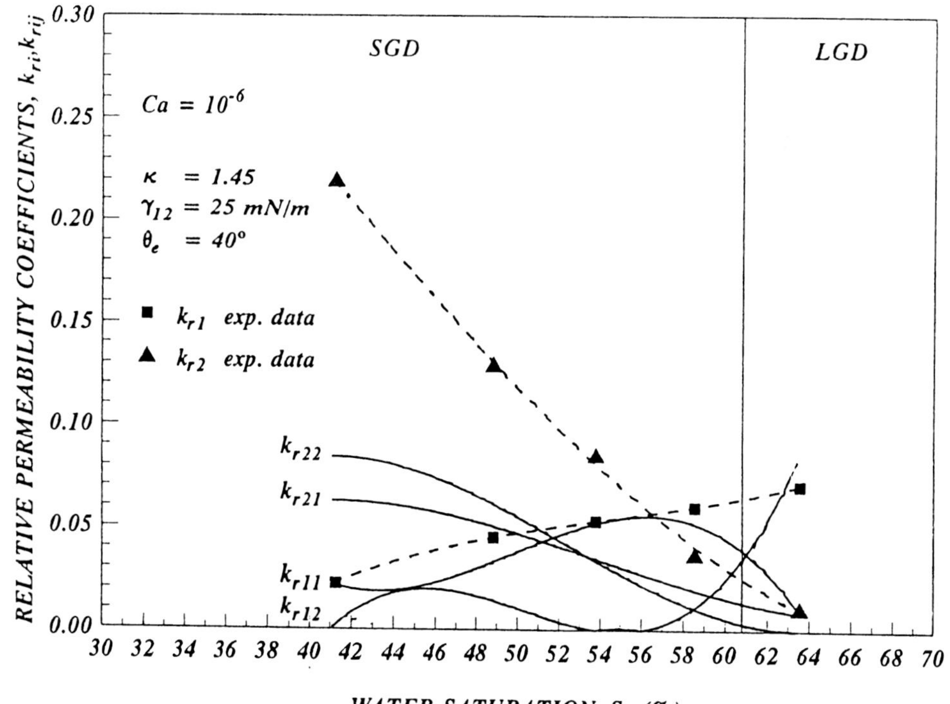

RELATIVE PERMEABILITY COEFFICIENTS, k_{ri}, k_{rij}
0.30
0.25
0.20
0.15
0.10
0.05
0.00
SGD
LGD
$Ca = 10^{-6}$
$\kappa = 1.45$
$\gamma_{12} = 25\ mN/m$
$\theta_e = 40°$
k_{r1} exp. data
k_{r2} exp. data
k_{r22}
k_{r21}
k_{r11}
k_{r12}
30 32 34 36 38 40 42 44 46 48 50 52 54 56 58 60 62 64 66 68 70
WATER SATURATION, S_1 (%)

Fig. 4b.

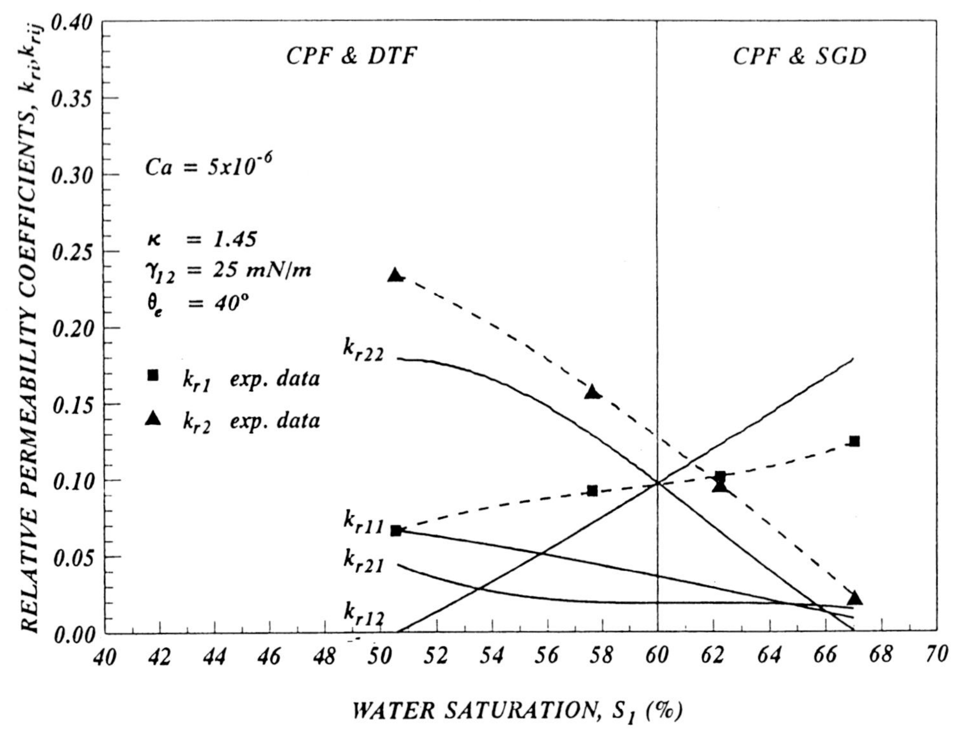

Fig. 4c.

Fig. 4. Conventional (dashed lines) and generalized (solid lines) relative permeability coefficients for $\kappa = 1.45$, with all the other physicochemical properties kept constant. (a) $Ca = 10^{-7}$, (b) $Ca = 10^{-6}$, (c) $Ca = 5 \times 10^{-6}$.

5.3. FLOW REGIMES AND COUPLING INDICES

The relative importance of the coupling effects can be expressed appropriately through the coupling indices, χ_i, which are introduced here as

$$\chi_i = 1 - \frac{k_{rii}}{k_{ri}}, \quad i = 1, 2. \tag{13}$$

The coupling index, χ_i, is the fraction of the flow rate of fluid i that is caused by coupling effects. An alternative expression for χ_i is

$$\chi_i = \frac{\mu_i}{\mu_j}\frac{k_{rij}}{k_{ri}}\left(\frac{\Delta P_j}{\Delta P_i}\right), \quad i, j = 1, 2, \quad i \neq j. \tag{14}$$

Equation (14) follows from a simple expression that relates the conventional and generalized coefficients, and which is obtained by combining Equations (1) and (2).

$$k_{ri} = k_{rii} + \frac{\mu_i}{\mu_j}k_{rij} + \frac{\mu_i}{\mu_j}k_{rij}\frac{\Delta P_j - \Delta P_i}{\Delta P_i} = k_{rii} + \frac{\mu_i}{\mu_j}k_{rij}\frac{\Delta P_j}{\Delta P_i},$$

$$i, j = 1, 2, \quad i \neq j. \tag{15}$$

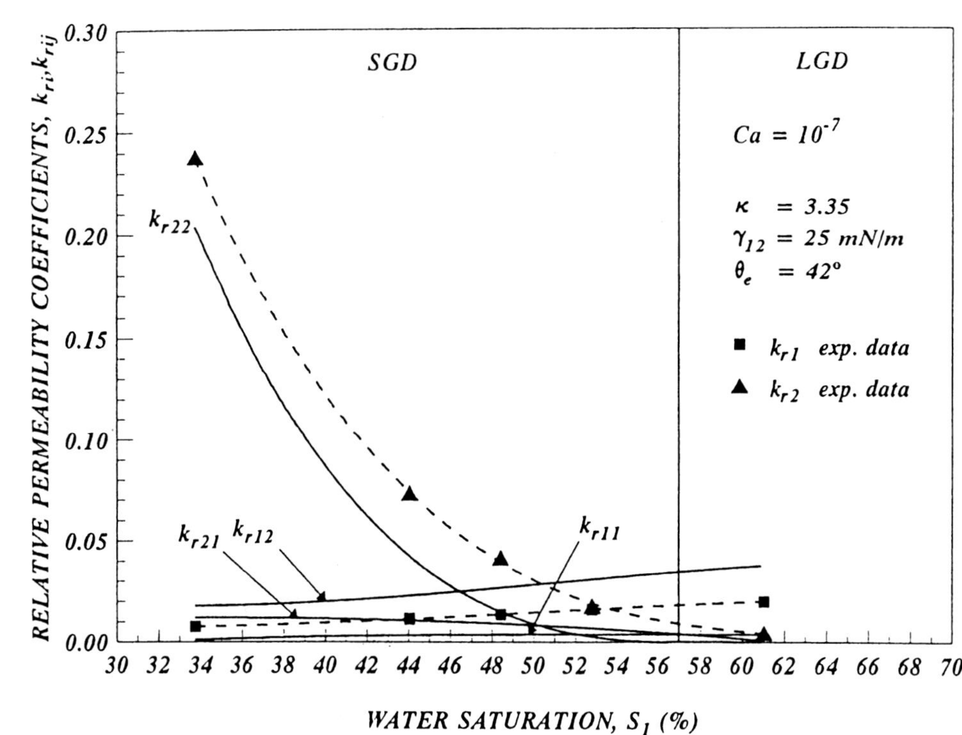
RELATIVE PERMEABILITY COEFFICIENTS, k_{ri}, k_{rij}
0.30
0.25
0.20
0.15
0.10
0.05
0.00
SGD
LGD
$Ca = 10^{-7}$
$\kappa = 3.35$
$\gamma_{12} = 25\ mN/m$
$\theta_e = 42°$
k_{r1} exp. data
k_{r2} exp. data
k_{r22}
k_{r21}
k_{r12}
k_{r11}
30 32 34 36 38 40 42 44 46 48 50 52 54 56 58 60 62 64 66 68 70
WATER SATURATION, S_1 (%)

Fig. 5a.

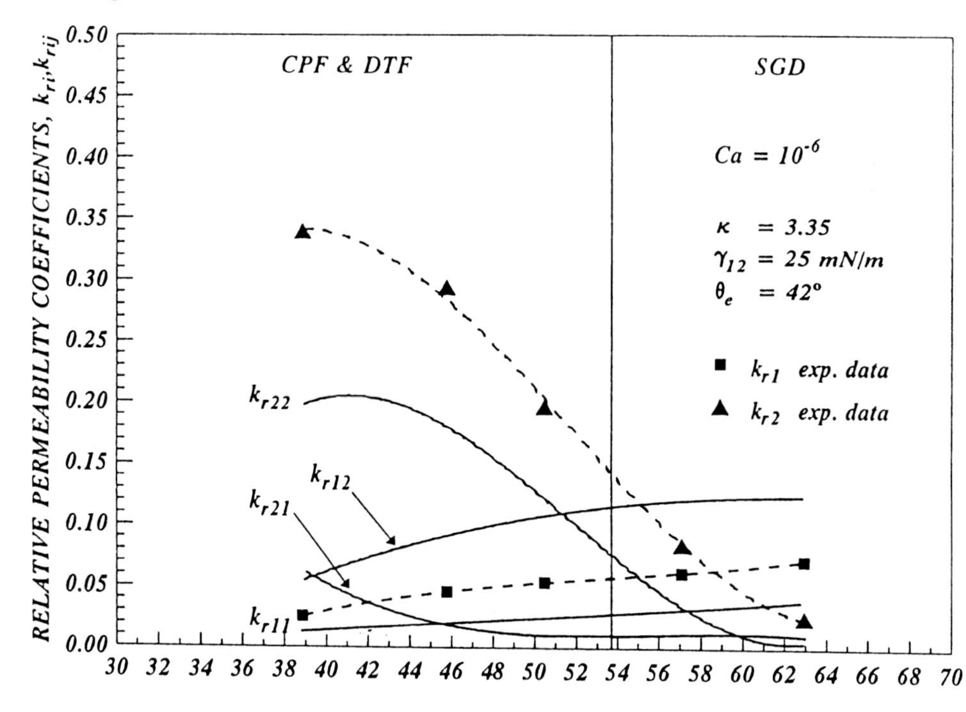
RELATIVE PERMEABILITY COEFFICIENTS, k_{ri}, k_{rij}
0.50
0.45
0.40
0.35
0.30
0.25
0.20
0.15
0.10
0.05
0.00
CPF & DTF
SGD
$Ca = 10^{-6}$
$\kappa = 3.35$
$\gamma_{12} = 25\ mN/m$
$\theta_e = 42°$
k_{r1} exp. data
k_{r2} exp. data
k_{r22}
k_{r12}
k_{r21}
k_{r11}
30 32 34 36 38 40 42 44 46 48 50 52 54 56 58 60 62 64 66 68 70
WATER SATURATION, S_1 (%)

Fig. 5b.

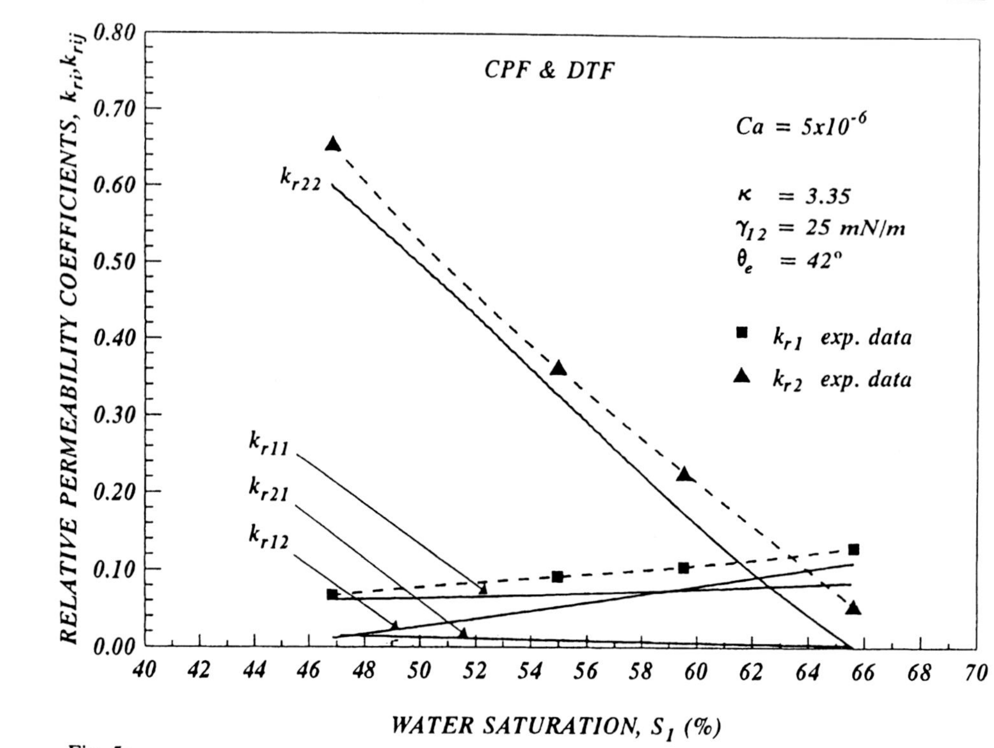

Fig. 5c.

Fig. 5. Conventional (dashed lines) and generalized (solid lines) relative permeability coefficients for $\kappa = 3.35$, with all the other physicochemical properties kept constant. (a) $Ca = 10^{-7}$, (b) $Ca = 10^{-6}$, (c) $Ca = 5 \times 10^{-6}$.

The combination of the two models is permissible here because the experimental data utilized by the method correspond not only to the same flow rates and pressure drops but also (a condition that is very important) to the same type of flow (steady-state, co-current), and the same flow regimes. For fully developed flow in an infinite porous medium, which is free from end and boundary effects, we have $\Delta P_i = \Delta P_j$, and Equations (15) and (14) reduce to

$$k^{\circ}_{ri} = k^{\circ}_{rii} + \frac{\mu_i}{\mu_j} k^{\circ}_{rij}, \quad i, j = 1, 2, \quad i \neq j \tag{16}$$

and

$$\chi^{\circ}_i = 1 - \frac{k^{\circ}_{rii}}{k^{\circ}_{ri}} = \frac{\mu_i}{\mu_j} \frac{k^{\circ}_{rij}}{k^{\circ}_{ri}}, \quad i, j = 1, 2, \quad i \neq j, \tag{17}$$

where the supersript ° denotes quantities that are free from end and boundary effects. Laboratory experiments that are entirely free from end-effects are very difficult to achieve. However, in the present work we take advantage of the small but discernible end effects and, thus, turn a small disadvantage to a desirable feature.

TABLE I. Application of Wald's criterion[a] for model discrimination

R =Ratio of Model 1 to Model 2 likelihood values	$Ca = 10^{-7}$	$Ca = 10^{-6}$	$Ca = 5 \times 10^{-6}$
$\kappa = 0.66$	1130	266	0.0413
$\kappa = 1.45$	19.8	6.49	0.5790
$\kappa = 3.35$	32.9	6.53	0.0018

Model 1: Conventional fractional flow model.
Model 2: Generalized fractional flow model.
At confidence level 95%:
- if $R \geqslant 19.0$ Model 1 is superior.
- if $R \leqslant 0.0526$ Model 2 is superior.
- if $0.0526 < R < 19.0$ more experiments are required to decide this issue.

[a]See Bard (1974).

It is this small difference ($\Delta P_1 \neq \Delta P_2$ but $|\Delta P_1 - \Delta P_2| \ll |\Delta P_1|$) that allows us to use Equation (15) rather than (16), and thus to determine the coupling effect contributions. When Equation (16) holds exactly, one cannot isolate the coupling effects from the overall behavior, because the RHS of Equation (16) functions as a single coefficient. This point requires elaboration.

It is important to understand that the coupling effects are *not* created by the end-effects. Rather, the small perturbation caused by the end-effects provides sufficient differentiation (ca. 10%) between the two pressure-drop signals to allow a statistically significant *decoupling* of the composite parameters $[k^{\circ}_{rii} + (\mu_i/\mu_j)k^{\circ}_{rij}]$. The weak end-effects under consideration do *not* significantly affect the details of the *flow mechanisms* in the region of the porous medium over which measurements were made. This was confirmed as follows. The region of measurements was partitioned in six segments, the flow behavior (number of ganglia per unit area, ganglion size distribution) was videorecorded and digitized and a statistical analysis was made to detect any significant differences between the time-averaged results among the six compartments. None were found. Thus, the coupling effects must be attributed to the flow mechanisms. It is also pertinent to point out that in the case of 'infinitely long' porous media sufficient perturbation to differentiate the two pressure gradient signals can be introduced by a *slight* local or (spatially periodic) variation of the mean pore diameter (or of the wettability) on a mesoscopic scale.

Figure 6 shows the application of both Equationss (16) and (15) to one of the cases in which the generalized coefficients have been estimated. Dashed lines present the results of Equation (16), whereas solid lines present those of Equation (15). The dotted lines are the estimated conventional relative permeabilities. Clearly, Equation (15) gives a sigificantly better approximation to the conventional

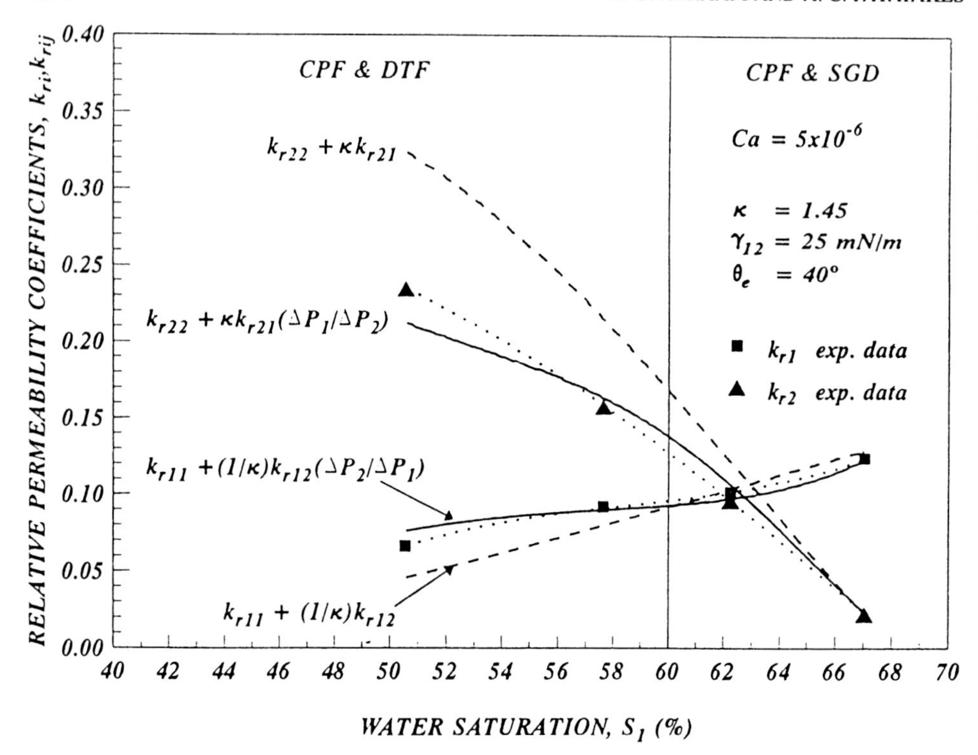

Fig. 6. Illustration of the simple relationship between the conventional and the generalized relative permeability coefficients for one set of flow conditions ($Ca = 5 \times 10^{-6}$, $\kappa = 1.45$). The estimated curves of the conventional relative permeability coefficients are presented with dotted lines.

permeabilities, because in these experiments there exist weak but discernible end effects, and $\Delta P_i \neq \Delta P_j$.

Figure 7 shows the coupling indices that were determined in this work and the map of the flow regimes, which was obtained in Avraam and Payatakes (1995). It should be kept in mind that because of the way in which the coupling coefficients, χ_i, are calculated from the relative permeability coefficients k_{ri} and k_{rii}, Equation (13), the relative error of χ_i is roughly equal to the sum of the relative errors of k_{ri} and k_{rii}. Consequently, the values of χ_1 and χ_2 shown in Figure 7, and tabulated in Table II, should be considered as rough approximations. Still, certain important conclusions can be drawn.

The coupling index of oil, χ_2, is large almost everywhere; it diminishes only near the domain of CPF, especially for $\kappa > 1$. In the domain of LGD and SGD, χ_2 is large (in many cases near unity), and it is, in general, an increasing function of S_1 (decreasing function of S_2). These results are strong evidence that the bodies of disconnected oil (ganglia, droplets) are driven to a large extent by the flowing water that engulfs them. However, the fact that χ_2 is usually less than unity indicates that ganglion-ganglion interactions make a sizable contribution. Such oil–oil interactions are observed frequently in our experiments. For example, when a moving ganglion collides with another slowly moving ganglion, the pair (even if

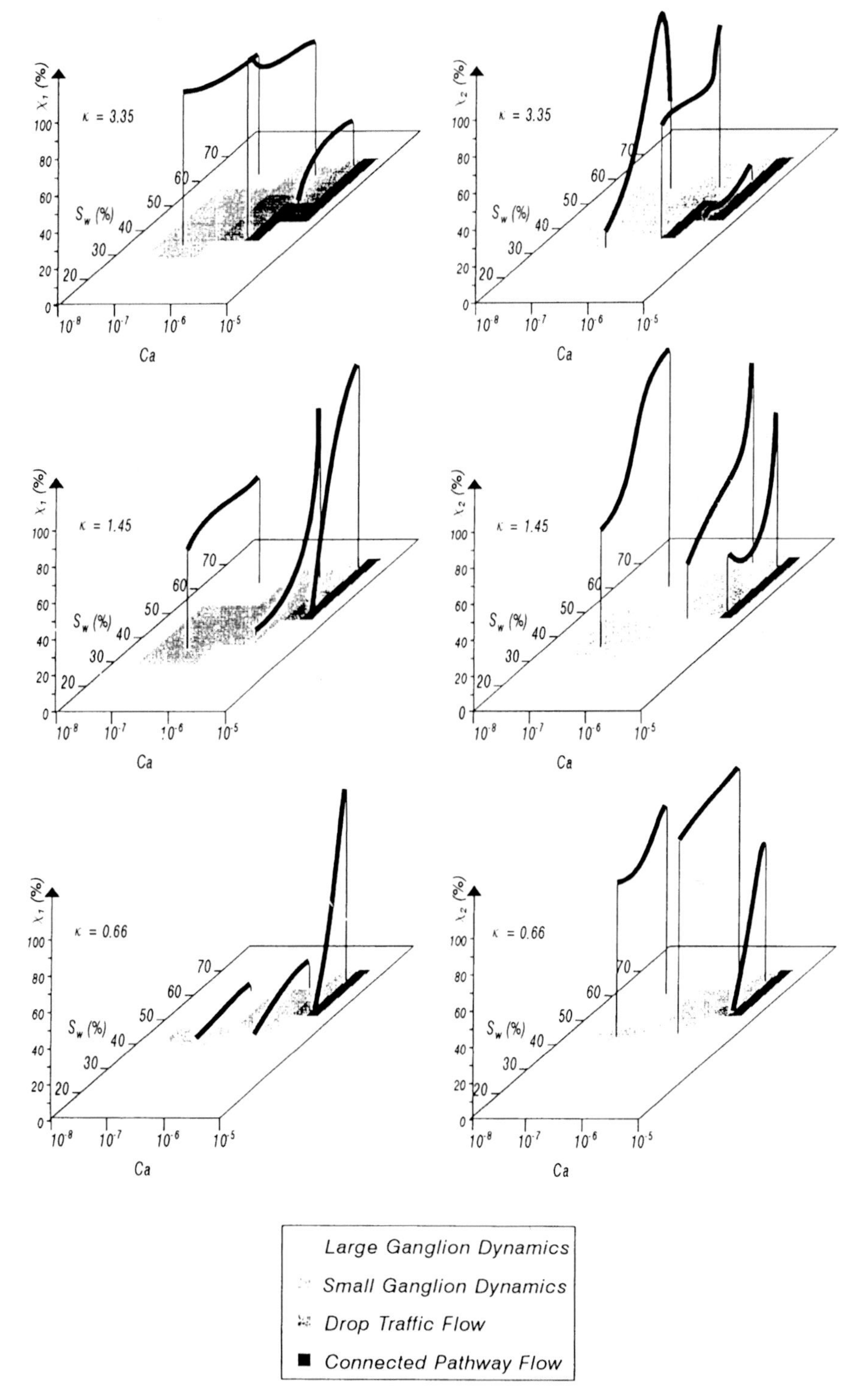

Fig. 7. Map of flow regimes and estimated coupling indices. The graph shows a strong correlation between the coupling indices and the flow regimes. For discussion see the text.

TABLE IIa. Estimated values of the conventional relative permeabilities, k_{r1}, k_{r2}, and the coupling indices, χ_1, χ_2

$Ca = 10^{-6}$

$\kappa = 3.35$

$S_1(\%)$	$k_{r1}(\%)$	$k_{r2}(\%)$	$\chi_1(\%)$	$\chi_2(\%)$
60.9	1.97	0.44	56.52	66.92
52.8	1.60	1.90	57.28	129.79
48.4	1.36	3.93	58.39	76.55
44.0	1.11	7.40	60.50	47.99
33.7	7.59	23.80	70.40	16.89

$\kappa = 1.45$

$S_1(\%)$	$k_{r1}(\%)$	$k_{r2}(\%)$	$\chi_1(\%)$	$\chi_2(\%)$
61.7	1.87	0.17	51.63	110.48
56.2	1.58	1.03	50.19	111.50
51.6	1.34	1.84	51.11	99.29
46.2	1.10	3.41	52.41	79.09
35.4	7.36	10.90	46.57	51.88

$\kappa = 0.66$

$S_1(\%)$	$k_{r1}(\%)$	$k_{r2}(\%)$	$\chi_1(\%)$	$\chi_2(\%)$
63.0	1.110	0.141	2.84	81.56
61.3	1.020	0.324	3.02	80.05
57.5	0.905	0.679	2.83	79.66
50.5	0.830	1.720	1.19	76.57
43.0	0.564	4.650	0.00	72.14

the ganglia do not coalesce) moves faster than the slower of the two ganglia before the collision. Another effect is the mobilization of stranded ganglia by oncoming moving ganglia.

The coupling index of water, χ_1, is also substantial in most cases, with some exceptions. In the domain near CPF and for $\kappa = 3.35$, χ_1 becomes small, especially for small values of S_1. In the domains of LGD and SGD, and for $\kappa < 1$, χ_1 is small, even nil. It is also evident that χ_1 is larger when $\kappa > 1$, than it is when $\kappa < 1$. This can be explained by the fact that the moving portions of oil exert larger viscous stresses on the water, as the oil viscosity increases, all other factors being the same. The peaks observed in the domain CPF and SGD for large S_1 (small S_2) values are attributed to the fact that in this flow regime a substantial amount of water is carried

TABLE IIb. Estimated values of the conventional relative permeabilities, k_{r1}, k_{r2}, and the coupling indices, χ_1, χ_2

$Ca = 10^{-6}$

$\kappa = 3.35$

$S_1(\%)$	$k_{r1}(\%)$	$k_{r2}(\%)$	$\chi_1(\%)$	$\chi_2(\%)$
62.9	7.19	2.42	53.38	86.40
57.1	6.21	8.55	55.44	59.92
50.5	5.37	20.50	58.73	61.08
45.8	4.59	28.50	63.30	59.93
38.9	2.54	34.10	66.73	59.33

$\kappa = 1.45$

$S_1(\%)$	$k_{r1}(\%)$	$k_{r2}(\%)$	$\chi_1(\%)$	$\chi_2(\%)$
63.5	7.04	1.03	81.50	96.99
58.5	6.02	4.14	32.61	68.99
53.8	5.31	8.16	16.34	62.37
48.8	4.50	13.20	11.76	55.69
41.2	2.29	21.90	2.19	41.33

$\kappa = 0.66$

$S_1(\%)$	$k_{r1}(\%)$	$k_{r2}(\%)$	$\chi_1(\%)$	$\chi_2(\%)$
64.2	6.26	1.10	11.84	83.40
61.6	5.48	1.90	12.19	80.24
58.8	4.84	3.24	11.68	79.85
53.2	3.97	7.23	8.24	78.32
44.0	2.10	17.20	0.00	72.14

along in the form of films and piston-like segments formed among neighboring small ganglia.

Application of Wald's criterion (Bard, 1974) for model discrimination (Table I) shows that Model 1 (without coupling terms) is superior to Model 2 (with coupling terms) in the domain of steady-state ganglion dynamics, at a confidence level of 95 %. In the domain of connected pathway flow the two models are virtually equivalent, at a confidence level of 95 %, with a slight (but not always statistically significant) superiority of Model 2.

6. Conclusions

A reliable and easy to use method for the determination of the generalized relative permeability coefficients from a set of steady-state co-current two-phase flow data

TABLE IIc. Estimated values of the conventional relative permeabilities, k_{r1}, k_{r2}, and the coupling indices, χ_1, χ_2

$Ca = 5 \times 10^{-6}$

$\kappa = 3.35$

$S_1(\%)$	$k_{r1}(\%)$	$k_{r2}(\%)$	$\chi_1(\%)$	$\chi_2(\%)$
65.5	13.00	5.70	25.37	10.32
59.5	10.50	23.00	27.47	2.65
55.0	9.22	36.10	24.35	0.26
46.8	6.71	65.50	8.93	2.78

$\kappa = 1.45$

$S_1(\%)$	$k_{r1}(\%)$	$k_{r2}(\%)$	$\chi_1(\%)$	$\chi_2(\%)$
67.0	12.40	2.32	99.56	92.50
62.3	10.20	9.55	83.16	28.55
57.6	9.10	15.90	53.51	17.69
50.6	6.63	23.40	0.00	27.39

$\kappa = 0.66$

$S_1(\%)$	$k_{r1}(\%)$	$k_{r2}(\%)$	$\chi_1(\%)$	$\chi_2(\%)$
68.0	11.30	1.61	93.92	63.28
65.2	9.02	3.55	60.98	38.84
61.6	7.71	6.61	25.74	14.28
57.6	6.69	10.20	0.00	0.00

was developed. It is a parameter estimation method that obtains the functional dependence of the generalized coefficients in a global way, by choosing the optimal functional representation of the unknown coefficients in order to match the experimental data with statistical rigor. The same method can also be used to obtain the conventional relative permeabilities. This method has the fundamental advantage that it uses data from flows that are of the *same nature* with that of the flow under consideration, namely, steady-state co-current two-phase flow. This is both convenient and imperative, since the nature of the flow affects every flow-related quantity, including the values of the relative permeabilities and the magnitude of the coupling effects.

The new method was used to analyze the flow data from the parametric experimental investigation of steady-state co-current two-phase flow that was reported in Avraam and Payatakes (1995). The main conclusions are the following.

- The viscous coupling effects are important over broad ranges of the values of the flow parameters S_1, Ca, κ, keeping all other factors (wettability, coalescence factor, porous medium) fixed.
- The cross coefficients in Equations (2a) and (2b) are found to be *unequal*, $k_{r12}/\mu_2 \neq k_{r21}/\mu_1$. For the class of flows under consideration this is *not* a violation of the Onsager–Casimir reciprocity relation, because the underlying flow mechanisms involve many *catastrophic* events (ganglion breakup, coalescence, stranding, mobilization), and therefore *do not possess microscopic reversibility.* (The Onsager–Casimir reciprocal relations hold only for phenomena that have microscopic reversibility.)
- The relative importance of the coupling effects is expressed appropriately through the *coupling indices*, which are introduced here, and which are defined as $\chi_i = 1 - (k_{rii}/k_{ri})$. These indices take values in the range (0, 1) and each of them is the normalized measure of the contribution of coupling effects to the flow rate of the corresponding fluid.
- The coupling indices are found to correlate strongly with the flow regimes, Fig. 7. These experimental results are in accord with the underlying flow mechanisms at pore level, and this provides a substantial insight into the process. It is established, among other things, that the bodies of disconnected oil (ganglia and droplets) are driven mainly by the flowing water that engulfs them. However, interactions between colliding bodies of oil are also significant. More detailed discussion of these issues is given in the main part of this work.
- A strikingly simple relationship is established between the generalized relative permeability coefficients and the conventional ones, specifically, Equation (15). In fully development flows (free of end and boundary effects) it is $\Delta P_j/\Delta P_i = 1$, and the above expression reduces to Equation (16). This relation explains the reason for which the coupling effects are so hard to isolate through macroscopic measurements; they are incorporated seamlessly in the values of k_{ri}. In the present work their isolation became possible because the data contained small but discernible differences between ΔP_1 and ΔP_2, caused by weak end-effects. *The coupling effects are inherent in the flow.* The weak end-effects made it possible to differentiate the two pressure drop signals, and thus to decouple the composite parameters $[k^{\circ}_{rii} + (\mu_i/\mu_j)k^{\circ}_{rij}]$ in a statistically significant way. [In 'infinitely long' porous media this can be achieved by introducing slight local perturbations in the mean pore diameter (or the wettability, etc.) on a mesoscopic scale.]
- Based on Wald's criterion for model discrimination (at a confidence level of 95 %) the conventional fractional flow model (no coupling effects) is superior to the generalized fractional flow model (with coupling effects) in the domain of steady-state ganglion dynamics. In the domain of connected pathway flow the two models are virtually equivalent, with a slight (but not always statistically sigificant) superiority of the generalized model. However, this is *not*

to be interpreted as evidence that the coupling effects are insignificant (see above).

- The present work suggests a synthesis. One can use the conventional relative permeability coefficients to describe the macroscopic behavior of steady-state co-current two-phase flow, but one should not lose sight of the fact that significant portions of their magnitudes, expressed by the coupling indices χ_i, are caused by coupling effects. (This simple approach does *not* apply to situations involving steep macroscopic gradients of the saturation and the capillary pressure, such as displacements fronts, etc.)
- The magnitudes of the coupling effects, and the flow mechanisms that produce them, are of fundamental importance for the development of a satisfactory (self-consistent) quantitative model or computer-aided simulator.

Acknowledgements

This work was supported by Shell Research B.V., Koninklijke/Shell, Exploratie en Produktie Laboratorium (KSEPL), and the Institute of Chemical Engineering and High Temperature Chemical Processes.

References

1. Amaefule, J. O., and Handy, L. L.: 1982, The effects of interfacial tensions on relative oil/water permeabilities of consolidated porous media, *Soc. Petrol. Eng. J.* June, 371–381.
2. Archer, J. S., and Wong, S. W.: 1973, Use of a reservoir simulator to interpret laboratory waterflood data, *Soc. Petrol. Eng. J.* Dec., 343–347.
3. Auriault, J. -L.: 1987, Nonsaturated deformable porous media: Quasastatics, *Transport in Porous Media* **2**, 45–64.
4. Auriault, J.-L., Lebaigue, O., and Bonnet, G.: 1989, Dynamics of two immiscible fluids flowing through deformable porous media, *Transport in Porous Media* **4**, 105–128.
5. Avraam, D. G., Kolonis, G. B., Roumeliotis, T. C., Constantinides, G. N., and Payatakes, A. C.: 1994, Steady-state two-phase flow through planar and non-planar model porous media *Transport in Porous Media* **16**, 75–101.
6. Avraam, D. G., and Payatakes, A. C.: 1995, Flow regimes and relative permeabilities during steady-state two-phase flow in porous media, *J. Fluid Mech.* **293**, 181–206.
7. Bard, Y.: 1974, *Non-Linear Parameter Estimation*, Academic Press, New York.
8. Batycky, J. P., McCaffery, F. G., Hodgins, P. K., and Fisher, D. B.: 1981, Interpreting relative permeability and wettability from unsteady-state displacement measurements, *Soc. Petrol. Eng.* June, 296–308.
9. Bentsen, R. G.: 1974, Conditions under which the capillary term may be neglected, *J. Canad. Petrol. Technol.*, Oct.–Dec., 25–30.
10. Bentsen, R. G., and Manai, A. A.: 1993, On the use of conventional cocurrent and countercurrent effective permeabilities to estimate the four generalized permeability coefficients which arise in coupled, two-phase flow, *Transport in Porous Media* **11**, 243–262.
11. Bourbiaux, B. J., and Kalaydjian, F.: 1988, Experimental study of cocurrent and countercurrent flows in natural porous media, Paper SPE 18283, *63rd Ann. Tech. Conf. and Exhibition of SPE*, Houston.
12. Bourbiaux, B. J., and Kalaydjian, F.: 1990, Experimental study of cocurrent and countercurrent flows in natural porous media, *SPERE* **5**, 361–368.
13. Buckley, S. E., and Leverett, M. C.: 1942, Mechanism of fluid displacement in sands, *Trans. AIME* **146**, 107–116.
14. Chatzis, J. D., Morrow, N. R., and Lim, H. T.: 1983, Magnitude and detailed structure of residual oil saturation, *Soc. Petrol. Eng. J.* **23**, 311–326.

15. Chen, J. D.: 1986, Some mechanisms of immiscible fluid displacement in small networks, *J. Coll. Int. Sci.* **110**, 488–503.
16. Constantinides, G. N., and Payatakes, A. C.: 1991, A theoretical model of collision and coalescence, *J. Coll. Int. Sci.* **141**(2), 486–504.
17. De Gennes, P. G.: 1983, Theory of slow biphasic flows in porous media, *Phys. Chem. Hydrodyn.* **4**, 175–185.
18. De la Cruz, V., and Spanos, T. J. T.: 1983, Mobilization of oil ganglia, *AIChE J.* **29**, 854–858.
19. Ehrlich, R.: 1993, Viscous coupling in two-phase flow in porous media and its effect on relative permeabilities, *Transport in Porous Media* **11**, 201–218.
20. Fulcher, R. A., Ertekin, T., and Stahl, C. D.: 1985, Effect of capillary number and its constituents on two-phase relative permeability measurements, *J. Petrol. Technol.*, Feb., 249–260.
21. Goode, P. A., and Ramakrishnan, T. S.: 1993, Momentum transfer across fluid-fluid interfaces in porous media: a network model, *AIChE J.* **39**, 1124–1134.
22. Heavyside, J., Black, C. J. J., and Berry, J. F.: 1983, Fundamentals of relative permeability: Experimental and theoretical considerations, paper SPE 12173, *58th Ann. Tech. Conf. Exhib.*, San Francisco, CA, October 5–8.
23. Honarpoor, M., and Mahmood, S. M.: 1988, Relative-permeability measurements: An overview, *J. Petrol. Technol.* Aug., 963–966.
24. Ioannidis, M. A., Chatzis, I., and Payatakes, A. C.: 1991, A mercury porosimeter for investigating capillary phenomena and microdisplacement mechanisms in capillary networks, *J. Coll. Int. Sci.* **143**, 22–36.
25. Jerault, G. R., and Salter, S. J.: 1990, The effect of pure structure on hysteresis in relaive permeability and capillary pressure: Pore-level modeling, *Transport in Porous Media* **5**, 103–151.
26. Johnson, E. F., Bossler, D. R., and Naumann, V. O.: 1959, Calculation of relative permeability from displacement experiments, *Trans. AIME* **216**, 370–372.
27. Jones, S. C., and Roszelle, W. O.: 1978, Graphical techniques for determining relative permeability from displacement experiments, *J. Petrol. Technol.* May, 807–817.
28. Kalaydjian, F.: 1987, A macroscopic description of multiphase flow in porous media involving evolution of fluid/fluid interface, *Transport in Porous Media* **2**, 537–552.
29. Kalaydjian, F., and Legait, B.: 1987a, Ecoulement lent a contre-courant de deux fluides non miscibles dans un capillaire présentant un rétrecissement, *C. R. Acad. Sci. Paris, Ser. II.* **304**, 869–872.
30. Kalaydjian, F., and Legait, B.: 1987b, Perméabilités relatives couplées dans des écoulement en capillaire et en milieu poreux, *C. R. Acad. Sci. Paris, Ser. II* **304**, 1035–1038.
31. Kalaydjian, F., Bourbiaux, B., and Guerillot, D.: 1989, Viscous coupling between fluid phases for two-phase flow in porous media: theory versus experiment, *Eur. Symp. on Improved Oil Recovery*, Budapest, Hungary.
32. Kalaydjian, F.: 1990, Origin and quantification of coupling between relative permeabilities for two-phase flows in porous media, *Transport in Porous Media* **5**, 215–229.
33. Kerig. P. D., and Watson, A. T.: 1986, Relative-permeability estimation from displacement experiments: An error analysis, *Soc. Petrol. Eng.* March, 175–182.
34. Lefebvre Du Prey, E. J.: 1973, Factors affecting liquid-liquid relative permeabilities of a consolidated porous medium, *Soc. Petrol. Eng. J.* Feb., 39–47.
35. Lelievre, R. F.: 1966, Etude d'écoulements diphasiques permanent a contre-courants en milieux–Comparison avec les écoulements de meme sens. Ph.D. Thesis University of Toulouse, France.
36. Lenormand, R., Zarcone, C., and Sarr, A.: 1983, Mechanisms of the displacement of one fluid by another in a network of capillary ducts, *J. Fluid Mech.* **135**, 337–355.
37. Leverett, M. C.: 1941, Capillary behavior in porous solids, *Trans AIME* **142**, 159–169.
38. Marle, C. M.: 1982, On macroscopic equations governing multiphase flow with diffusion and chemical reactions in porous media, *Int. J. Eng. Sci.* **20**, 643–662.
39. McCaffery, F. G., and Bennion, D. W.: 1974, The effect of wettability on two-phase relative permeabilities, *J. Canad. Petrol. Technol.* Oct.–Dec., 42–53.
40. Morrow, N. R., and McCaffery, F. G.: 1978, in G. F. Padday (ed), *Wetting, Spreading, and Adhesion*, Academic Press, New York.

41. Naar, J., Wygal, G. R., and Henderson, J. H.: 1962, Imbibition relative permeability in unconsolidated porous media, *Soc. Petr. Eng. J.* **2**, 13–23.
42. Odeh, A. S.: 1959, Effect of viscosity ratio on relative permeability, *J. Petrol. Technol.* **11**, 346–354.
43. Owens, W. W., and Archer, D. L.: 1971, The effect of rock wettability on oil–water relative permeability relationships, *J. Petrol. Technol.* July, 873–878.
44. Raats, P. A. C., and Klute, A.: 1968, Transport in soils: The balance of momentum, *Soil Sci. Soc. Am. J.* **32**, 452–166.
45. Rapoport, L. A., and Leas, W. J.: 1953, Properties of linear waterfloods, *Trans. AIME*, **198**, 139–148.
46. Richards, L. A.: 1931, Capillary conduction of liquids through porous mediums, *Physics*, **1**, 318–333.
47. Rose, W.: 1972, *Fundamentals of Transport Phenomena in Porous Media*, IAHR, Elsevier, New York.
48. Rose, W.: 1988, Measuring transport coefficients necessary for the description of coupled two-phase flow of immiscible fluids in porous media, *Transport in Porous Media* **3**, 163–171.
49. Rose, W.: 1989, Data interpretation problems to be expected in the study of coupled two-phase flow of immiscible fluid flows in porous media, *Transport in Porous Media* **4**, 185–198.
50. Rose, W.: 1991, Richards' assumptions and Hassler's presumptions, *Transport in Porous Media* **6**, 91–99.
51. Sanchez-Palencia, E.: 1974, Compartiment local et macroscopique d'un type de milieux physiques et heterogenes, *Int. J. Engn. Sci.* **12**, 331–351.
52. Sandberg, C. R., Gournay, L. S., and Sippel, R. F.: 1958, The effect of fluid-flow rate and viscosity on laboratory determinations of oil-water relative permeabilities, *Trans. AIME* **213**, 36–43.
53. Sigmund, P. M., and McCaffery, F. G.: 1979, An improved unsteady-state procedure for determining the relative permeability characteristics of heterogeneous porous media, *Soc. Petrol. Eng. J.* Feb., 15–28.
54. Spanos, T. J. T., de la Cruz, V., Hube, J., and Sharma, R. C.: 1986, An analysis of Buckley–Leverett theory, *J. Can. Petr. Tech.* **25**(1), 71–75.
55. Taber, J. J.: 1958, The injection of detergent slugs in water floods, *Trans. AIME* **213**, 186–192.
56. Tao, T. M., and Watson, A. T.: 1984, Accuracy of JBN estimates of relative permeability, Part 1, Error analysis, *Soc. Petrol. Eng.* Apr., 215–224.
57. Vizika, O., and Payatakes, A. C.: 1989, Parametric experimental study of forced imbibition in porous media, *Phys. Chem. Hydrodyn.* **11**, 187–204.
58. Wardlaw, N. C.: 1982, The effects of geometry, wettability, viscosity and interfacial tension on trapping in single pore-throat pairs, *J. Canad. Petrol. Technol.* **21**, 21–27.
59. Watson, A. T., Richmond, P. C., Kerig, P. D., and Tao, T. M.: 1988, A regression-based method for estimating relative permeabilities from displacement experiments, *SPERE*, Aug., 953–958.
60. Welge, H. L.: 1952, A simplified method for computing oil recovery by gas or water drive, *Trans. AIME*, **195**, 91–98.
61. Whitaker. S.: 1986, Flow in porous media, II. The governing equations for immiscible two-phase flow, *Transport in Porous Media*, **1**, 105–125.
62. Wright, R. J., and Dawe, R. A.: 1980, An examination of the multiphase Darcy model of fluid displacement in porous media, *Rev. Inst. Franç. Petrole* Nov.–Dec. **XXXV**(6), 1011–1024.
63. Yadav, G. D., Dullien, F. A. L., Chatzis, I., and McDonald, I. F.: 1987, Microscopic distribution of wetting and nonwetting phases in sandstones during immiscible displacement, *SPERE* **2**, 137–147.
64. Yortsos, Y. C., and Fokas, A. S.: 1983, An analytical solution for linear waterflood including the effects of capillary pressure, *Soc Petrol. Eng. J.* **23**, 115–124.
65. Yuster, S. T.: 1951, Theoretical considerations of multiphase flow in idealized capillary systems, *World Petroleum Cong. Proc., Section II. Drilling and Production*, The Hague.

Transport in Porous Media **20:** 169–196, 1995.

Effective Properties for Flow Calculations

M. J. KING, P. R. KING, C. A. McGILL and J. K. WILLIAMS
BP Exploration Operating Company Ltd., Chertsey Road, Sunbury-on-Thames, Middlesex, TW16 7LN, U.K.

(Received: May 1994)

Abstract. In this paper we discuss the background to the problems of finding effective flow properties when moving from a detailed representation of reservoir geology to a coarse gridded model required for reservoir performance simulation. In so doing we synthesize the pictures of permeability and transmissibility and show how they may be used to capture the effects of the boundary conditions on the upscaling. These same concepts are applied to the renormalization method of calculating permeability, to show its promise as an accurate, yet fast method.

Key words: effective flow properties, reservoir geology, permeability, transmissibility.

1. Introduction

Reservoir descriptions are invariably generated on a fine scale, in part reflecting the scale of the input information (such as core data). This is done in the belief that it is necessary to capture as much of the geological heterogeneity as possible for accurate prediction of fluid flow. Reservoir performance simulators tend to be on a much coarser scale to enable manageable computation. Consequently, the determination of a faithful coarse scale representation of the aggregated effects of smaller scale heterogeneities is a key problem in reservoir characterisation.

Reservoir performance simulators generally use finite difference schemes for numerical flow modelling. The reservoir volume is divided into grid blocks which are then assigned values for the reservoir parameters (porosities, absolute permeabilities, transmissibilities, relative permeabilities, and so on). As a result, reservoir rock properties are implicitly considered to be homogeneous on the scale of the grid blocks used. However, these properties must reproduce the effects of heterogeneities on all scales. These coarse scale parameters are known as effective grid block properties. The definition of effective properties is not entirely unambiguous and there are many in the literature. We shall follow the definition of the effective permeability and/or transmissibility as that which would give the same total flow through an 'equivalent' homogeneous region as the flow through the heterogeneous region for the same boundary conditions. We recommend a specific sequence of boundary conditions for these calculations and show how the cell permeability and the face transmissibilities encode information about the effects of the boundary conditions.

For nonadditive properties like permeability the determination of the effective property is not straightforward. The main difficulty lies in the interdependent influences of heterogeneities on many length scales. Many different upscaling procedures have been proposed [1–26]. All involve assumptions and approximations of one form or another. No single procedure is completely general, and many are problem-specific.

A major component in the variability of reservoir performance prediction arises from our incomplete knowledge of the given subsurface environment and the inherent uncertainty in the spatial distribution of petrophysical properties. Stochastic modelling tools provide a means for assessing the impact of this uncertainty [1–10]. The idea is to generate a series of random realizations of the reservoir consistent with the available data. Subsequent calculations with a reservoir performance simulator allow one to gain an appreciation of the impact of this uncertainty.

A key intermediate step in this process is upscaling. This step introduces its own additional uncertainties due to the approximations made, in particular assumptions about boundary conditions. The question of boundary conditions will be taken up again later. We turn now to a consideration of the desirable features of a technique for upscaling under uncertainty.

Upscaling methods should be reliable, accurate, adaptable, and efficient. Some compromises are inevitable, and depend on the problem at hand. In the face of considerable uncertainty, what is wanted is not the upscaled behaviour for any one realization but the distribution of upscaled behaviour over an ensemble of realizations. 'Exact' methods may give too much precision for any realization, at great computational expense, and encounter the statistical fluctuation problem in the ensemble averaging, too few realizations can be processed to characterize the ensemble fully. 'Approximate' methods provide less precision for any one realization, but do so quite cheaply, allowing a large number of realizations to be processed. So a good approximation method needs to be fast and sufficiently accurate to permit processing of sufficient realizations to obtain statistically meaningful results encompassing the full range of uncertainty and to avoid any serious systematic bias.

We know how to solve (numerically) many upscaling problems very precisely (in principle, if not in practice), yet we do not know how to solve them less precisely and cheaply. In many cases, it is difficult to compute (accurate) low-precision answers reliably. We have no idea of the size of the error until we have computed a very accurate answer.

As alluded to above, a major difficulty arises from the vexed question of boundary conditions. When determining the effective permeability of a given grid block, it is not possible to anticipate the actual flow boundary conditions that block will experience during a full-field simulation. These boundary conditions may well be both space- and time-dependent (for multiphase flow). In the development that follows the implications of the choice of boundary conditions on the mathematical formulation is made explicit.

A common practice is to solve the flow equation with no flow boundary conditions applied to all of the faces of the block except for two opposite faces that are set to constant, but different, pressures. While this approach appears eminently reasonable, it is not clear what errors are incurred.

In this paper we shall discuss some of the above issues. We shall discuss what is meant by upscaling and some of the requirements of an 'effective' property. We shall discuss the errors associated with the calculation of these properties and indicate some ways of estimating or reducing these errors. In this context we shall use the term 'effective' property to apply to an appropriate property used in a numerical simulator. This need not necessarily be the same as that determined from physical principals alone as discussed in the next section.

2. Permeability–Differential Formulation

For a region with constant pressure gradient, ∇P, and constant velocity, $\mathbf{V}$, we may define permeability as the coefficient, K, in Darcy's Law: $\mathbf{V} = -\mathsf{K} \bullet \nabla P$. This simple starting point becomes rapidly complicated when one attempts to implement it in an upscaling calculation, where the implications are quite different depending upon whether Darcy's Law is being used as a differential equation (local property) or within a finite difference/finite element calculation. In the latter case, the implementation depends upon the shape of the element (rectangular, triangular, or PEBI), and the order of the pressure/velocity representations we utilize during the upscaling. We will discuss the numerical formulation, covering permeability and the closely related effective property, transmissibility, in the next section, while covering the properties of absolute effective permeability for the differential form of Darcy's Law now.

Much discussion has centred on the nature of the permeability tensor and its symmetry [12, 15, 23–26]. There are two arguments that lead directly to a symmetric tensor – that based on energy dissipation and one based on Onsager's relationships. Essentially these arguments are the same and hold for a continuum description of the problem.

The Onsager argument follows as a consequence of the fluctuation dissipation theorem that argues that the way a fluctuation builds up is the same as the way it decays. So consider a fluid in a porous medium subject to a pressure gradient ∇P. The local rate of dissipation of energy (per unit volume) is $-\mathbf{V} \bullet \nabla P$. If one asserts a linear constitutive relationship between velocity and pressure gradient (i.e., flows are small and hence laminar) $V_i - K_{ij}\partial_j P$ then reversal of the pressure field yields reversal of the streamlines and the energy dissipation (and hence entropy production is unaltered) if and only if the permeability tensor is symmetric.

In energy dissipation terms the only contribution comes from the symmetric part of the tensor. The argument is as follows [26]. The energy dissipation in a given volume is $I = \int -V_i\partial_i P\,\mathrm{d}\tau$ with the auxiliary relation $V_i = -K_{ij}\partial_j P$ which can be considered as a constraint to the above and so introducing Lagrange multipliers we have

$$I = \int [-V_i\partial_i P + \lambda_i(V_i + K_{ij}\partial_j P)]\,\mathrm{d}\tau$$

which must be minimized with respect to variations in V and P. Let $P \rightarrow P + \varepsilon\theta$ and $V_i \rightarrow V_i + \varepsilon u_i$. Then

$$\begin{aligned}\delta I &= \int [-u_i\partial_i P - V_i\partial_i\theta + \lambda_i(u_i + K_{ij}\partial_j\theta)]\,\mathrm{d}\tau \\ &= \int [-u_i(\partial_i P - \lambda_i) + \theta\partial_i(V_i - \lambda_j K_{ji})]\,\mathrm{d}\tau + \text{surface terms.}\end{aligned}$$

Requiring this to be zero for all possible variations in θ and u yields $\lambda_i = \partial_i P$ and $\partial_i(V_i - \lambda_j K_{ji}) = 0$. Substituting for the Lagrange multiplier and the original constitutive law gives the conservation equation $\partial_i(K_{ij} + K_{ji})\partial_j P = 0$. This is consistent with the original supposition if and only if the permeability is symmetric.

So both these arguments lead to the conclusion that permeability is described by a symmetric tensor when the flow is written in a continuum form. Do these arguments apply when we choose to write a finite difference approximation to the flow? In general not. We will show this explicitly in Section 5, but can discuss the physics in general terms first. Consider a block of material that we will homogenize and then represent with a finite difference grid block. If the boundary conditions (say pressure) are prescribed everywhere over the surface of that block then we may solve the flow equations based on volume conservation and Darcy's Law ($\nabla \bullet (\mathsf{K} \bullet \nabla P) = 0$) to obtain ∇P and $\mathbf{V}$ interior to the block. However, neither ∇P, $\mathbf{V}$, nor $\mathbf{V} \bullet \nabla P$ are constant, and almost always $\langle \mathbf{V} \bullet \nabla P\rangle \neq \langle \mathbf{V}\rangle \bullet \langle \nabla P\rangle$. In other words we may no longer evaluate the dissipation of the system in terms of the homogenized properties ($\langle \nabla P\rangle$ and $\langle \mathbf{V}\rangle$): we have lost access to the variational formulation and the symmetry properties it provides.

There is a second issue associated with the observation that ∇P is not generally constant. Since this assumption is key to the definition of permeability, it raises the question of whether permeability is suitable for parameterizing a numerical representation of flow. These issues, will be discussed in Section 5, as will the dependence of the answer on the particular elements of the numerical calculation (shape of element and order of the scheme).

3. Finite Difference Calculations and Upscaling

In this section we review the flow picture which underlies the standard finite difference scheme. Several generalizations will become obvious, but their discussion will be deferred until Section 5. By the end of this section we will be working with cell permeabilities, face transmissibilities, and upscaling algorithms that capture the range of ambiguity associated with the boundary conditions of the flow. For ease of formulation we work in a finite-element language and for pedagogic reasons we introduce the additional complications of a general corner-point cell. We start the section by reviewing the corner-point cell geometry.

3.1. CORNER-POINT CELL GEOMETRY

Consider the general corner-point cell [27]. The cell is defined by a tri-linear interpolation in $(\alpha, \beta, \gamma) \in [-\frac{1}{2}, \frac{1}{2}]$ between the eight corner points:

$$\begin{aligned}\mathbf{x} = & \left(\tfrac{1}{2}+\gamma\right)\left(\left(\tfrac{1}{2}+\beta\right)\left(\left(\tfrac{1}{2}+\alpha\right)\mathbf{x}_{\mathrm{TNE}} + \left(\tfrac{1}{2}-\alpha\right)\mathbf{x}_{\mathrm{TNW}}\right) + \right.\\ & \left. +\left(\tfrac{1}{2}-\beta\right)\left(\left(\tfrac{1}{2}+\alpha\right)\mathbf{x}_{\mathrm{TSE}} + \left(\tfrac{1}{2}-\alpha\right)\mathbf{x}_{\mathrm{TSW}}\right)\right) + \\ & +\left(\tfrac{1}{2}-\gamma\right)\left(\left(\tfrac{1}{2}+\beta\right)\left(\left(\tfrac{1}{2}+\alpha\right)\mathbf{x}_{\mathrm{BNE}} + \left(\tfrac{1}{2}-\alpha\right)\mathbf{x}_{\mathrm{BNW}}\right) + \right.\\ & \left. +\left(\tfrac{1}{2}-\beta\right)\left(\left(\tfrac{1}{2}+\alpha\right)\mathbf{x}_{\mathrm{BSE}} + \left(\tfrac{1}{2}-\alpha\right)\mathbf{x}_{\mathrm{BSW}}\right)\right).\end{aligned} \tag{3.1}$$

The face directions are labelled by East (E) and West (W), North (N) and South (S), and Top (T) and Bottom (B), and the corner points accordingly. There are three tangent vectors which characterize the cell:

$$\mathbf{t}_{10} = \left.\frac{\partial \mathbf{x}}{\partial \alpha}\right|_{(0,0,0)}, \qquad \mathbf{t}_{20} = \left.\frac{\partial \mathbf{x}}{\partial \beta}\right|_{(0,0,0)}, \qquad \mathbf{t}_{30} = \left.\frac{\partial \mathbf{x}}{\partial \gamma}\right|_{(0,0,0)}, \tag{3.2}$$

evaluated at cell centre. Each tangent vector may be visualized as extending from a face centre to the opposite face centre. The corresponding unit vectors are

$$\hat{t}_{\ell 0} \equiv \mathbf{t}_{\ell 0}/|\mathbf{t}_{\ell 0}|, \quad l = 1,\ 2,\ 3. \tag{3.3}$$

For the evaluation of flux and transmissibility, we also need to know the (local) normal planes and directions, i.e., the vector orthogonal to a pair of tangent vectors that lie in the plane.

$$\mathbf{n}_1 = \mathbf{t}_2 \times \mathbf{t}_3, \quad \mathbf{n}_2 = \mathbf{t}_3 \times \mathbf{t}_1, \quad \mathbf{n}_3 = \mathbf{t}_1 \times \mathbf{t}_2. \tag{3.4}$$

Because the corner-point cell is more general than a parallelepiped, those normals vary algebraically with position through the cell. For instance, $\mathbf{n}_1$ varies quadratically in α and bi-linearly in β and in γ. The faces of the corner-point cell are *not* in general planes, instead they appear to be twisted ribbons and are known as ruled surfaces. Figure 1 is a sketch of a corner-point cell showing block-centred tangents and face normals.

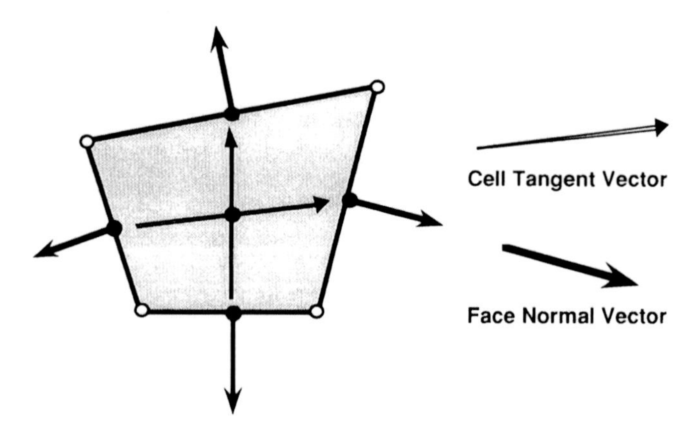

Fig. 1. Corner-point cell.

Volumes are determined by the integral of the Jacobian of the transformation from (α, β, γ) to (x, y, z):

$$\int\int_V\int \mathrm{d}x\,\mathrm{d}y\,\mathrm{d}z = \int\int_V\int \mathrm{d}\alpha\,\mathrm{d}\beta\,\mathrm{d}\gamma \left|\frac{\partial(x,y,z)}{\partial(\alpha,\beta,\gamma)}\right|, \tag{3.5}$$

where the Jacobian is a polynomial in (α, β, γ):

$$\text{Jacobian} = \left|\frac{\partial(x,y,z)}{\partial(\alpha,\beta,\gamma)}\right| = \begin{vmatrix} \dfrac{\partial x}{\partial\alpha} & \dfrac{\partial x}{\partial\beta}, & \dfrac{\partial x}{\partial\gamma} \\ \dfrac{\partial y}{\partial\alpha} & \dfrac{\partial y}{\partial\beta} & \dfrac{\partial y}{\partial\gamma} \\ \dfrac{\partial z}{\partial\alpha} & \dfrac{\partial z}{\partial\beta} & \dfrac{\partial z}{\partial\gamma} \end{vmatrix} \tag{3.6}$$

and hence is easy to evaluate. We use a two-point Gaussian quadrature to integrate the Jacobian in each coordinate (α, β, γ) to determine the cell volume.

3.2. UPSCALING TENSOR PERMEABILITY

As discussed in the previous section, for a homogeneous porous media a constant pressure gradient generates a constant velocity, related by the permeability tensor K, via Darcy's Law:

$$\mathbf{V} = -\mathsf{K} \bullet \nabla P \tag{3.7}$$

To upscale a region of *heterogeneous* porous media we impose a constant value of ∇P on the boundary of the region, perform a finite difference calculation to evaluate the fine cell pressures and velocities, and then the volume averaged velocity, $\langle \mathbf{V} \rangle$, to define the

$$\langle \mathbf{V} \rangle = -\mathsf{K}_{\text{eff}} \bullet \nabla P \tag{3.8}$$

With three independent choices of ∇P, and three components for $\langle \mathbf{V} \rangle$, we obtain a 3×3 effective permeability tensor.

The averaged velocity is defined as the volume weighted velocity for each cell, j:

$$\langle \mathbf{V} \rangle = \sum_j \text{Vol}_j \cdot \mathbf{V}_j \Big/ \sum_j \text{Vol}_j \,, \tag{3.9}$$

where $\mathbf{V}_j$ is defined as the constant vector which gives the 'best' representation to the fluxes though the six faces, $f = 1, \ldots, N_f$, of the cell. Because $\mathbf{V}$ need not be constant in a cell, we minimize $\chi^2 = \frac{1}{2} \sum_{f=1}^{N_f} (Q_f - \mathbf{n}_f \bullet \mathbf{V})^2$, being the mis-match of the estimate of flux through each face, to determine the best choice of constant velocity. Minimization of χ^2 with respect to $\mathbf{V}$ leads to the matrix equation for $\mathbf{V}$.

$$\left(\sum_{f=1}^{N_f} \mathbf{n}_f \mathbf{n}_f \right) \bullet \mathbf{V} = \sum_{f=1}^{N_f} \mathbf{n}_f Q_f, \tag{3.10}$$

where the sum is over the faces of the cell.

3.3. VECTOR PERMEABILITY

In practice we often work with a simplified three-component form of the permeability tensor, the so-called permeability vector:

$$\mathsf{K} = \sum_\ell K_\ell \hat{t}_{\ell 0} \hat{t}_{\ell 0}. \tag{3.11}$$

Once we have determined the effective permeability tensor we may calculate the permeability vector $\mathbf{K} = \{K_\ell | \ell = 1, 2, 3\}$ be requiring that the velocities predicted by the more general model agree with those constructed from the simpler model. In particular, for each K_ℓ we have

$$\mathbf{V} = -K_\ell \hat{t}_{\ell 0} \hat{t}_{\ell 0} \bullet \nabla P \tag{3.12}$$

from the simple model and Equation (3.7) for the full model. Components of ∇P orthogonal to $\hat{t}_{\ell 0}$, which do not contribute to Equation (3.12), must be chosen in Equation (3.7) to generate a velocity parallel to $\hat{t}_{\ell 0}$. Explicitly, $0 = \{\mathsf{K} - K_\ell \hat{t}_{\ell 0} \hat{t}_{\ell 0}\} \bullet \nabla P$, or recognising that this must be true for nontrivial ∇P, one obtains that the determinant, $||\mathsf{K} - K_\ell \hat{t}_{\ell 0}\hat{t}_{\ell 0}||$, must vanish. Working in the $\{\hat{n}_{\ell 0}\}$ coordinate system gives

$$0 = ||\hat{n}_{i0} \bullet \mathsf{K} \bullet \hat{n}_{j0} - K_\ell \delta_{il}\delta_{jl}(\hat{n}_{\ell 0} \bullet \hat{t}_{\ell 0})^2|| \tag{3.13}$$

or

$$K_\ell = \frac{1}{(\hat{n}_{\ell 0} \bullet \hat{t}_{\ell 0})^2} \frac{||\hat{n}_{i0} \bullet \mathsf{K} \bullet \hat{n}_{j0}||}{||C_\ell[\hat{n}_{i0} \bullet \mathsf{K} \bullet \hat{n}_{j0}]||}, \tag{3.14}$$

where $C_\ell[\hat{n}_{i0} \bullet \mathsf{K} \bullet \hat{n}_{j0}]$ is the matrix co-factor obtained by dropping row and column ℓ.

3.4. UPSCALING VECTOR PERMEABILITY

The upscaling calculation follows the same approach as already outlined in reducing K to K, i.e., the general choice of boundary conditions are restricted so that the velocity is aligned with $\hat{t}_{\ell 0}, \ell = 1, 2, 3$. Because we now have access to velocities along the boundary we can satisfy this requirement both locally and on the average. Locally, we reduce the transmissibility across each fine cell face on the boundary according to its orientation:

$$T_{\text{edge},k} = T_k \cdot |\hat{n}_k \bullet \hat{t}_{l0}|. \tag{3.15}$$

The reason for the factor of $\hat{n}_k$ will become apparent when transmissibility is discussed. Globally, we utilize three independent boundary conditions, $\nabla P = \{-\hat{t}_{\ell 0}, -\hat{n}_{j0} | j \neq \ell\}$ and superpose them with weights

$$\{t_{\ell 0}, t_{\ell 0} c_j | j \neq \ell\} : \nabla P = -t_{\ell 0} \left\{ \hat{t}_{\ell 0} + \sum_{j \neq \ell} c_j \hat{n}_{j0} \right\},$$

to calculate $\langle \mathbf{V} \rangle = \{\langle \mathbf{V} \rangle_\ell + \sum_{j \neq \ell} c_j \langle \mathbf{V} \rangle_j\}$. Choosing $\{c_j | j \neq \ell\}$ to satisfy $\hat{n}_{j0} \bullet \langle \mathbf{V} \rangle = 0, j \neq \ell$, fixes the upscaling calculation, leading finally to

$$K_\ell = \mathbf{n}_{\ell 0} \bullet \langle \mathbf{V} \rangle|_{\nabla P = -\mathbf{t}_{\ell 0}} / (\mathbf{n}_{\ell 0} \bullet \mathbf{t}_{\ell 0}). \tag{3.16}$$

Physically, we drive fluid from face to face (the pressure drop is along $-\mathbf{t}_{\ell 0}$) and extract the component that moves fluid across a plane through the center of the cell ($\mathbf{n}_{\ell 0}$). This construction is summarized in Figure 2.

For the special case of rectangular cells, Equation (3.15) ensures that $\hat{n}_{j0} \bullet \langle \mathbf{V} \rangle = 0, j \neq \ell$, without the use of cross pressure terms to maintain the average flow direction. The standard algorithm for K_{eff} [13, 14] is a calculation based on restricted velocity and only requires a single upscaling calculation per flow direction. Otherwise, there will be three upscaling calculations per K_ℓ.

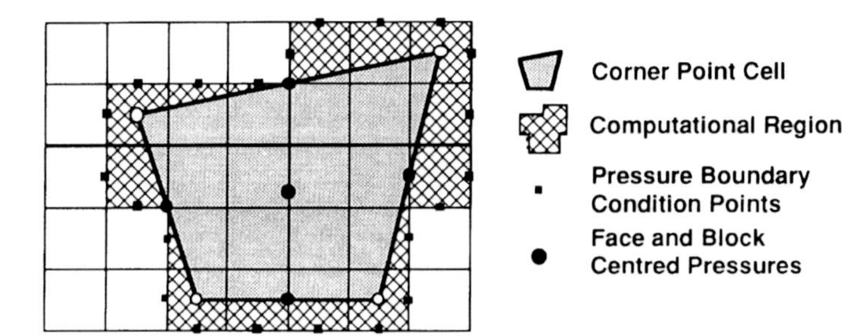

Fig. 2. Upscaling for permeability.

The resulting estimate of K_ℓ will always be less than the estimate obtained from K because it has a more restricted set of boundary conditions, Equation (3.15). Physically, the boundary conditions imposed through the full tensor allow locally leaky-side upscaling boundaries, so long as the average flux vanishes, while those that include transmissibility reduction will completely seal the side boundaries during the upscaling.

3.5. TRANSMISSIBILITY

Although we speak of upscaling effective permeability, the standard finite difference scheme is not based on cell permeability but instead on face transmissibility. How do we reduce the multiple degrees of freedom represented by K, to a single transport coefficient, T_f for each face $f = 1, \ldots, 6$? As in the reduction from K to $\mathbf{K}$, we do so by considering a restricted velocity field, and ask that the two representations agree. The construction follows by imposing a velocity field uniform and normal to the cell face.

$$\mathbf{V} = V\hat{n}_f \tag{3.17}$$

This construction does not require that the physical flow be normal to the face, but it does recognise that components of flow tangential to the face cannot be resolved by a single transport coefficient – the transmissibility. If there is a need to improve the resolution of cross-flow, then a more general finite difference scheme would have to be employed.

The magnitude of $\mathbf{V}$ is known in terms of T_f

$$Vn_f = Q_f = T_f \Delta P, \tag{3.18}$$

where

$$\begin{aligned} \Delta P &= -(\mathbf{x}_f - \mathbf{x}_0) \bullet \nabla P \\ &= -\tfrac{1}{2}\mathbf{t}_{f0} \bullet \nabla P. \end{aligned} \tag{3.19}$$

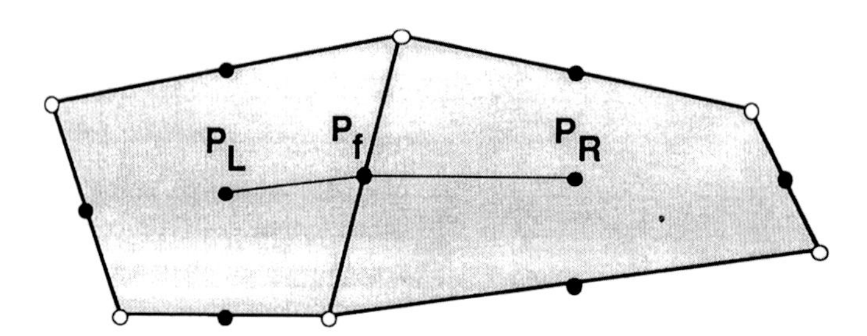

Fig. 3. Transmissibility construction.

Hence

$$\begin{aligned}\mathbf{V} &= -\mathbf{n}_f \frac{T_f}{2n_f^2}\mathbf{t}_{f0} \bullet \nabla P \\ &= -\mathbf{n}_\ell \frac{T_f}{2n_\ell^2}\mathbf{t}_{\ell 0} \bullet \nabla P,\end{aligned} \tag{3.20}$$

where we have re-indexed from the outwardly directed faces $f = 1,\ldots,6$ to directions $\ell = 1, 2, 3, 1, 2, 3$. However, all position-dependent quantities are evaluated at the centre of face f. Proceeding as in the derivation of Equation (3.13), the determinant $||\mathsf{K} - \mathbf{n}_\ell(T_f/2n_\ell^2)\mathbf{t}_{\ell 0}||$ must vanish which gives

$$T_f = \frac{(2n_l^2)}{(\mathbf{t}_l \bullet \mathbf{n}_l)(\mathbf{t}_{l0} \bullet \mathbf{n}_{l0}} \frac{||\mathbf{t}_i \bullet \mathsf{K} \bullet \mathbf{n}_{j0}||}{||C_\ell[\mathbf{t}_i \bullet \mathsf{K} \bullet \mathbf{n}_{j0}]||}. \tag{3.21}$$

For the reduced permeability model, Equation (3.11), we can replace Equation (3.21) with a more explicit construction. The determinant that now must vanish is (for $j = 1, 2, 3$)$||\mathbf{t}_{j0} - \delta_{j\ell}\mathbf{n}_\ell(T_f/2K_\ell n_\ell^2)t_{\ell 0}^2||$. Hence,

$$T_f = \frac{(2K_\ell n_\ell^2)}{(\mathbf{t}_l \bullet \mathbf{n}_\ell)(t_{l0}^2)} \frac{||\mathbf{t}_i \bullet \mathbf{t}_{j0}||}{||C_\ell[\mathbf{t}_i \bullet \mathbf{t}_{j0}]||}, \tag{3.22}$$

which shows that the face transmissibility can be written as the product of the permeability and a purely geometric factor.

The flux, Equation (3.18), relies on knowing the difference between face- and block-centre pressures. We can remove reference to the face pressure by invoking a transmissibility relation for the flux from the left side and from the right (Figure 3)

$$Q_f = T_L(P_L - P_f) = T_R(P_f - P_R). \tag{3.23}$$

Removing reference to P_f gives $(1/T_L + 1/T_R)Q_f = (P_L - P_R)$, hence $1/T = 1/T_L + 1/T_R$ or

$$T = T_L T_R/(T_L + T_R). \tag{3.24}$$

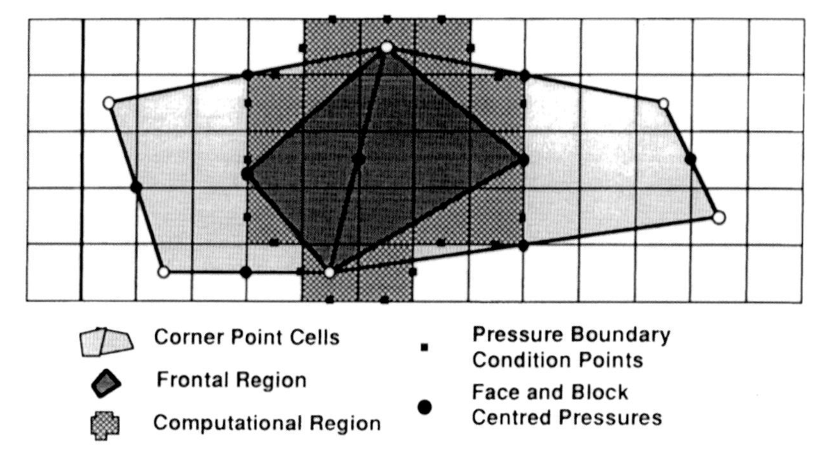

Fig. 4. Upscaling for transmissibility.

This harmonic sum is the corner-point version of the standard harmonic average construction for transmissibility [28]. The factor of two between harmonic average and sum was blended into the pre-factors in Equations (3.21), (3.22).

3.6. UPSCALING TRANSMISSIBILITY

Upscaling for transmissibility is very similar to upscaling for permeability as we still impose a pressure boundary condition on a computational region, restrict the velocity as before, and calculate the resulting flux using a finite difference calculation (typically based on the transmissibilities calculated by Equations (3.22), (3.24)). The restriction for velocity is with respect to the face normal Equation (3.17) giving

$$T_{\text{edge},k} = T_k \cdot |\hat{n}_k \bullet \hat{n}_f| \tag{3.25}$$

and the computational region is shifted to centre on the face (Figure 4). Two new features need to be introduced. The averaging region for the flux is only a subset of the computational region, and the pointwise presure drop, $(P_L - P_R)$, needs to be evaluated.

The velocity is averaged only over those cells that overlap the coarse cell face, as before evaluated using Equation (3.9). Flux is then obtained by $Q_f = \mathbf{n}_f \bullet \mathbf{V}$. To determine a pointwise pressure value within a cell, we stay with the constant gradient model:

$$P(\mathbf{x}) = P_C + (\mathbf{x} - \mathbf{x}_C) \bullet \nabla P, \tag{3.26}$$

where P_C is known directly and P_f indirectly (Equation (3.23)) from the finite difference calculation. We minimize $\chi^2 = \frac{1}{2}\sum_{f_1}^{N_f}\{P_C + (\mathbf{x}_f - \mathbf{x}_C) \bullet \nabla P - P_f\}^2$ to evaluate ∇P from the cell face pressures. The matrix equation is

$$\left(\sum_{f=1}^{N_f} \mathbf{t}_{f0}\mathbf{t}_{f0}\right) \bullet \nabla P = 2\sum_{f=1}^{N_f} \mathbf{t}_{f0}(P_f - P_C), \tag{3.27}$$

or recognizing the symmetry of the faces,

$$\left(\sum_{\ell=1}^{N} \mathbf{t}_{\ell 0}\mathbf{t}_{\ell 0}\right) \bullet \nabla P = 2\sum_{f=1}^{N_f} \mathbf{t}_{f0}P_f \tag{3.28}$$

Finally, this gives $\nabla P = (P_L - P_R)$ and

$$T_f\langle \mathbf{V}\rangle \bullet \mathbf{n}_f/(P_L - P_R) \tag{3.29}$$

for the effective transmissibility.

If elements are to be recombined in the future, then it is desirable to calculate the two transmissibilities that contribute to T_f by determining the face pressure and then defining T_L and T_R.

$$\begin{aligned} \frac{1}{T_f} &= \frac{(P_L - P_R)}{Q_f} = \frac{(P_L - P_f)}{Q_f} + \frac{(P_f - P_R)}{Q_f} \\ &= \frac{1}{T_L} + \frac{1}{T_R} \end{aligned} \tag{3.30}$$

This algorithm is sufficiently general to upscale flow in arbitrarily shaped elements. As shown in Figure 4, when upscaling transmissibility, we have reresolved the volume of the model from cell-centred elements to face-centred elements. General elements, such as the Perpendicular Equal-Area Bisector cells [29], can be resolved in this same fashion, generating finite difference codes where the number of bands is identical to the number of cell faces. Because the construction emphasizes the face it also captures hybrid cell structures that might be PEBI in some region and regular in others.

3.7. DISCUSSION

We now have three independent means of evaluating effective transmissibility. In the first we upscale to tensor permeability and then use Equation (3.21). Alternatively, we may restrict the flow according to the directions $\{\hat{t}_{\ell 0}|\ell = 1, 2, 3\}$, obtain the permeability vector, and use Equation (3.22). Finally, we may upscale for transmissibility, flow restricted by $\{\hat{n}_f|f = 1, \ldots, 6\}$, and use Equation (3.29). How do we choose between these possible algorithms? The choice comes to your understanding of flow in the vicinity of the averaging region, *which is not information available to any local averaging procedure*. If the choice of boundary conditions are consistent with the large-scale flow directions then the quality of the upscaling approximation will improve. However, if these directions are not known, we are still in a position to bound the answer and, hence, estimate the sensitivity of the result to changes in boundary conditions.

In general it makes sense to determine the effective transmissibility using both Equations (3.21) and (3.29), since the unrestricted flow will provide an upper estimate of transmissibility, while restricted flow will provide a lower one. The ratio of the two is stored as a transmissibility modifier; we expect the actual value to lie somewhere between this value of the modifier and unity.

If the modifier is significantly less than unity, then the user must supply a judgement on the correct choice of boundary conditions. One means of generating this judgement is to incrementally increase the size of the computational domain. As the edge blocks for which we impose Equation (3.25) are moved further from the region of interest, then the two computations must converge to the same answer. An explicit example of this is shown in the next section.

Upscaling to vector permeability and then computing transmissibility has limited utility in finite difference calculations. It does not supply an upper bound on transmissibility; this bound is provided by the tensor permeability upscaling which uses a more relaxed set of boundary conditions. Neither is permeability the property used directly in finite difference calculations. Nonetheless, it does supply a connection with Darcy's Law, and is a common point for understanding fluid flow in porous media. For this reason it is the most common flow characteristic upscaled. However, for utility in finite difference calculations we advocate upscaling for the two bounds instead.

This section is summarised in Tables I and II. Table I(a) lists the models of permeability and transmissibility while Table II lists the equation numbers for upscaling and for converting one type of effective property to another.

4. Error Analysis

So far we have considered the principles behind calculating and estimating effective properties and the associated errors. We have shown that these stem primarily from the use of inappropriate boundary conditions. We shall now examine this issue in more detail for a particular estimation technique, renormalization. The renormalization methed [18, 19, 21] is a hierarchical approach to calculating the effective permeability. The fine grid is consecutively coarse grained onto coarser grids by sequentially homogenising regions of the fine grid. This is done by asserting some boundary conditions on the equivalent coarse (renormalization) cells. For isotropic systems this turns out to be a surprisingly robust and accurate approach [20] with errors often less than 10%. However, for anisotropic systems the errors

TABLE Ia. Standard Upscaling Algorithms – based on a constant pressure gradient

	Standard models, constant ∇P					
Element	Differential		Rectangular			PEBI
Symbol	K	K	K	K	T	T
Type	Local	Local	Cell	Cell	Face	Face
Number	1	N^2	N	N^2	$2N$	N_f
Comment	Scalar	Symmetric Tensor $N \times (N+1)/2$ independent K's	Vector	Tensor		N_f faces

TABLE Ib. Generalized Upscaling Algorithms – based on 'complete' pressure boundary conditions

		Generalised models, complete (variable) ∇P			
Element		Triangular		Rectangular	PEBI
Symbol	K	T	T	T	T
Type	Cell	Cell	Face	Cell	Cell
Number	N^2	$(N+1)^2$	$(N+1)$	$(2N)^2$	N_f^2
Comment	Symmetric Tensor $N \times (N+1)/2$ independent K's	Equivalent to K $N \times (N+1)/2$ independent T's	Subset of K	$(2N-1)^2$ independent T's dependent upon scheme	N_f faces, $(N_f-1)^2$ independent T's

Element: differential, rectangular, triangular, or PEBI
Symbol: K = permeability, T = Transmissibility
Type: Local, cell, or face
Number: N = dimensionality, N_f = number of faces

TABLE II. Summary of equation numbers for effective properties

Rectangular cell		Convert to →			
Convert from	Upscaling equation	K Tensor	K Vector	Face T	Cell T
K tensor	(3.8)	–	(3.14)	(3.21)	(5.23)
K vector	(3.16)	(3.11)	–	(3.22)	(5.23)
Face T	(3.29)	(3.8)	(3.16)	–	(5.26)
Cell T	(5.7)	(5.24)	(5.25)	(5.26)	–

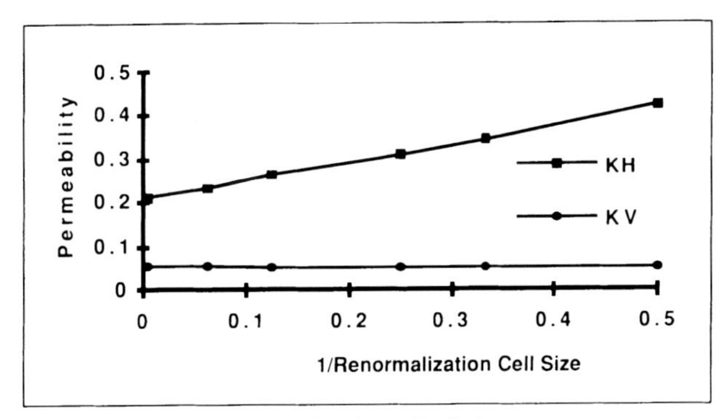

Fig. 5. Error in renormalization calculation.

can be considerable. We shall discuss this and indicate some resolution to this problem.

A detailed review of the renormalization method is not necessary as there are many in the literature [18–21]. However, whilst the literature demonstrates that the method can be surprisingly accurate there is little discussion of the errors. A detailed mathematical derivation of the errors was given for a simplified renormalization model in reference [30]. This demonstated that the principal source of the errors was the boundary conditions chosen for the simple case. There are two choices to improve the accuracy of the method. One is to improve the boundary conditions for the small cell. This may not be easy as small (2×2) cells are all boundary and so the results become overly sensitive to the boundary conditions. The other alternative is to use larger renormalization cells. This becomes quite attractive as the larger cells are still small for numerical calculation and the flow can be calculated using a direct numerical technique and, hence, very rapidly and accurately. This enables a sequence of calculations to be done at varying renormalization cell size. These results can then be extrapolated to infinite size, thereby recovering an accurate estimate of the 'true' results.

We shall demonstrate this using a simple example case. Consider a 256×256 grid. Each cell is assigned a horizontal permeability of 1 and a vertical permeability of 0.1. At random 20% of the sites are given zero permeability in each direction. Detailed simulation shows that the horizontal and vertical effective permeabilities are 0.210 and 0.0533 (with a standard error of ± 8 in the final place) respectively. However, the standard (18) 2×2 cell renormalization calculation gives $K_H = 0.422\,(18)$ and $K_V = 0.0497\,(9)$ (the figures in parentheses indicate the standard error over a number of realizations). This 100% error in the horizontal effective permeability is not atypical for the 2×2 cell renormalization for anisotropic systems. We then ran the calculation at a variety of other cell sizes, Table III.

It can clearly be seen that the accuracy of the renormalization result improves dramatically as the unit cell size increases. Plotting effective permeability against the reciprocal cell size (Figure 5) one sees a good linear relationship enabling

TABLE III. Influence of renormalization cell size on effective permeability calculation

Cell size	K_H	K_V	K_V/K_H
256	0.210(8)	0.0533(8)	0.25
16	0.233(9)	0.0523(9)	0.22
8	0.260(12)	0.0508(8)	0.20
4	0.307(15)	0.0494(8)	0.16
2	0.343(15)	0.0492(8)	0.14
2	0.422(18)	0.0497(9)	0.12

one to extrapolate from small cell results ($2 \times 2, 3 \times 3, 4 \times 4$) to the large system result. This linear relationship is expected as the reciprocal cell size is essentially the proportion of the small cell sites that are on the surface and, hence, contribute to the boundary-condition error. We would, therefore recommend that an estimate of the renormalization error can be made by estimating the slope of the linear relationship (e.g. Figure 5) by running a few small cell renormalization calculations.

5. Generalized Upscaling Formulation

The previous two sections have worked from Equation (3.8) and its variations: a constant ∇P determines a constant (or average) $\mathbf{V}$ with a transport coefficient K relating one to the other. We have varied the choice of boundary conditions (Section 3) and the resolution of those boundary conditions (renormalization cell size – Section 4), but retained this basic structure. In this section we recognize that the basic assumption is *false* – pressure generally varies in a more complex manner, even in the simplest of finite difference schemes. This observation is developed into a generalized upscaling formulation which provides a 'complete' description of the pressure variation (up to the order of the numerical scheme), a 'self-consistent' usage of that pressure representation to derive the boundary conditions, and a generalized linear response model to represent the upscaled properties. We believe that a consistent usage of this approach will remove the impact of the far field boundaries on the 'local' flow properties observed by many researchers [20, 25, 30].

5.1. TRIANGULAR ELEMENTS

There is one example of this generalized approach in the literature – upscaling to triangular elements [15, 23, 24]. We will discuss the upscaling in the standard manner first and then re-interpret the results in the more general form. Following Equation

(3.8) and Figure 2, we place a coarse triangular element upon a finer numerical mesh, and construct constant ∇P boundary conditions on the fine mesh,

$$P(\mathbf{x}) = P_C + (\mathbf{x} - \mathbf{x}_C) \bullet \nabla P. \tag{5.1}$$

Evaluating $\langle \mathbf{V} \rangle$ numerically gives K after N independent choices of ∇P.

First we observe that this pressure model is 'complete'. There are $N+1$ degrees of freedom in the element – the face pressures of the triangular cells, and as many degrees of freedom in the constant ∇P model, Equation (5.1) (N in the gradient, and one in the cell centre pressure). Proceeding as in the derivation of Equation (3.27), but including a variation with respect to P_C, we obtain Equation (3.27) and

$$P_C = \frac{1}{N_f} \sum_{f=1}^{N_f} P_f \tag{5.2}$$

after defining the cell centre and tangent vectors as

$$\mathbf{x}_C = \frac{1}{N_f} \sum_{f=1}^{N_f} \mathbf{x}_f, \tag{5.3}$$

$$\mathbf{t}_{f0} = 2(\mathbf{x}_f - \mathbf{x}_C), \tag{5.4}$$

where $\mathbf{x}_f$ is the face-centred location of pressure P_f. Solving Equation (3.27) gives

$$\nabla P = 2 \left\{ \sum_{f=1}^{N_f} \mathbf{t}_{f0} \mathbf{t}_{f0} \right\}^{-1} \bullet \sum_{f=1}^{N_f} \mathbf{t}_{f0} P_f \tag{5.5}$$

and finally for the flux through each face

$$Q_r = -2\mathbf{n}_r \bullet \mathsf{K} \bullet \left\{ \sum_{f=1}^{N_f} \mathbf{t}_{f0} \mathbf{t}_{f0} \right\}^{-1} \bullet \sum_{f=1}^{N_f} \mathbf{t}_{f0} P_f \tag{5.6}$$

or, in the most general form,

$$Q_r = \sum_{s=1}^{N_f} T_{rs}(P_s - P_C). \tag{5.7}$$

This is the most general linear response model, and replaces Equation (3.8) as the basis of the upscaling calculations.

Two restrictions will naturally arise. First, since volume (mass) must be conserved ($\sum_{r=1}^{N_f} Q_r = 0$) for all pressure drops, then the row sum vanishes:

$$\sum_{r=1}^{N_f} T_{rs} = 0. \tag{5.8}$$

Second, incompressibility will provide a linear relationship between the centre pressure and the face pressures, such as Equation (5.2),

$$P_C = \sum_{f=1}^{N_f} w_f P_f, \quad \sum_{f=1}^{N_f} w_f = 1. \tag{5.9}$$

The particular set of $\{w_f\}$ will depend upon the element and the pressure model. Substituting for P_C gives

$$Q_r = \sum_{s=1}^{N_f} \tilde{T}_{rs} P_s, \tag{5.10}$$

where

$$\tilde{T}_{rs} = T_{rs} - w_s \sum_{t=1}^{N_f} T_{rt}. \tag{5.11}$$

These modified transport coefficients have both vanishing column and row sums, with only $(N_f - 1)^2$ independent entries.

5.2. RECTANGULAR CELL PRESSURE MODELS

Can we repeat this approach for a rectangular cell? Not immediately, since there are $2N$ face pressures but still only $N + 1$ degrees of freedom in the constant ∇P model. It is necessary to generalise the pressure model first and then to repeat the approach. We present three pressures models, of increasing continuity, and describe their use as a basis for an upscaling calculation (Figure 6). Each model has $2N + 1$ pressures ($2N$ face centred pressures and one block centre pressure), and we will associate a permeability K with the element.

The simplest pressure model is obtained by breaking the element into $2N$ regions – one associated with each face. This model underlies the face transmissibility construction of Section 3, in other words, within each of the $2N$ regions ∇P is constant and aligned according to $\nabla P = -V\mathsf{K}^{-1} \bullet \hat{n}_f$, leading to the pressure model

$$P(\mathbf{x}) = P_C + (P_f - P_C)\frac{(\mathbf{x} - \mathbf{x}_c) \bullet \mathsf{K}^{-1} \bullet \mathbf{n}_f}{(2\mathbf{t}_{f0} \bullet \mathsf{K}^{-1} \bullet \mathbf{n}_f)}. \tag{5.12}$$

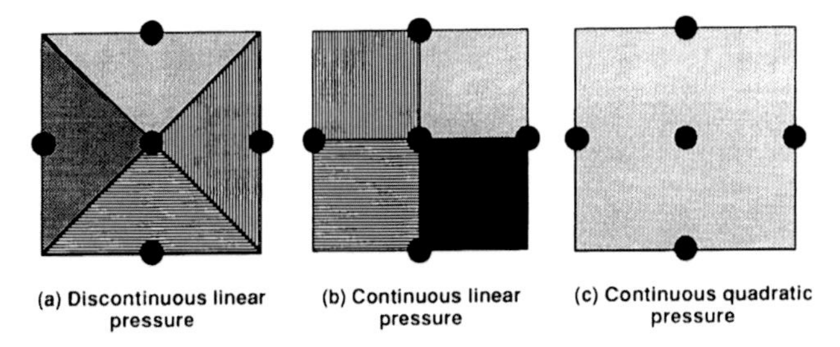

Fig. 6. Spatial support of three pressure models.

Can this function be used as a replacement for Equation (5.1) when fixing the upscaling pressure boundary conditions? Formally, the answer is yes, but physically the answer is no. If we superpose the coarse cell of Figure 6a on a fine grid we impose dis-continuous pressure values between each of the regions, leading to extremely high local pressure gradients and large local velocities. The resulting calculation has little information about the upscaling region – it is dominated by the corners of the region. Hence, we need more spatial continuity in the underlying pressure model if we are to use it as a basis for an upscaling calculation.

The next simplest model, but with a higher degree of spatial continuity, is shown in Figure 6b. The cell is split into quadrants (octants in three dimensions) with a pressure gradient specified by Equation (3.27). The pressure model of

$$P(\mathbf{x}) = P_C + 2(\mathbf{x} - \mathbf{x}_C) \bullet \left\{ \sum_{\ell=1}^{N} \mathbf{t}_{\ell 0} \mathbf{t}_{\ell 0} \right\}^{-1} \bullet \sum_{f=1}^{N} \mathbf{t}_{f0} (P_f - P_C), \tag{5.13}$$

where the sum is over the N face pressures in each quadrant, is different in each quadrant but predicts continuous values for $P(\mathbf{x})$ along the quadrant boundaries. As each of the $2N$ pressure differences are set to unity in turn,

$$(P_f - P_C) = \delta_{sf}, \quad s, f = 1, \ldots, 2N, \tag{5.14}$$

we determine the boundary conditions for a finite difference calculation. The results of the calulation are the $2N$ faces fluxes, Q_r determining the coefficients T_{rs}. Because the calculation is based on incompressible Darcy flow, Equation (5.8) will be automatically satisfied.

Within a single coarse cell, pressure is continuous at the internal quadrant boundaries but the local Darcy velocity need not be. If one requires a higher order model with continuous velocities, for instance for streamline or particle tracking applications, then we are lead to a quadratic model of pressure. Working with three directions, $\{\hat{e}_n | n = 1, 2, 3\}$ we can generate the coordinates:

$$x_n = (\mathbf{x} - \mathbf{x}_C) \bullet \hat{e}_n, \quad n = 1, 2, 3. \tag{5.15}$$

in which the pressure is quadratic

$$P(\mathbf{x}) = P_C + \sum_{n=1}^{N} (x_n P_n + \tfrac{1}{2} x_n^2 P_{nn}) \tag{5.16}$$

and the velocity linear

$$\mathbf{V} = -\sum_{n=1}^{N} \mathsf{K} \bullet \hat{e}_n [P_n + x_n P_{nn}] \tag{5.17}$$

as follows from Darcy's Law. Here P_n and P_{nn} are the first and second derivatives of pressure in the $\hat{e}_n$ direction. Typically, we choose the directions $\{\hat{e}_n\}$ to be the eigenvectors of the symmetric part of K, which is expected to minimise grid orientation effects. If K is not known, then we choose $\{\hat{e}_n\}$ w to be the cell tangent directions, $\{\hat{t}_{n0}\}$. These derivatives have the same number of degrees of freedom as the face to block pressure differences $P_f - P_C$, and so supply a complete representation.

The gradients are related to the face pressures by minimizing

$$\begin{aligned} \chi^2 &= \frac{1}{2} \sum_{f=1}^{N_f} \left\{ P_C + \sum_{n=1}^{N} (x_{nf} P_n + \tfrac{1}{2} x_{nf}^2 P_{nn}) - P_f \right\}^2, \\ x_{nf} &= (\mathbf{x}_f - \mathbf{x}_C) \bullet \hat{e}_n. \end{aligned} \tag{5.18}$$

The equations for $\{P_i, P_{ii} | i = 1, \ldots, N\}$ are

$$\begin{bmatrix} \sum_{\ell=1}^{N} (\hat{e}_i \bullet \mathbf{t}_{\ell 0}) \sum_{n=1}^{N} (\hat{e}_n \bullet \mathbf{t}_{\ell 0}) & 0 \\ 0 & \sum_{\ell=1}^{N} (\hat{e}_i \bullet \mathbf{t}_{\ell 0})^2 \sum_{n=1}^{N} (\hat{e}_n \bullet \mathbf{t}_{\ell 0})^2 \end{bmatrix} \begin{bmatrix} P_n \\ P_{nn} \end{bmatrix}$$

$$= \begin{bmatrix} \sum_{f=1}^{N_f} \hat{e}_i \bullet \mathbf{t}_{f0} P_f \\ 4 \sum_{f=1}^{N_f} (\hat{e}_i \bullet \mathbf{t}_{f0})^2 P_f - P_C) \end{bmatrix} \tag{5.19}$$

where the symmetries of the corner-point cell have been used to recognize that some terms vanish. The equation of flux continuity, $\nabla \bullet \mathbf{V} = 0$, will be satisfied identically if the second-order derivatives satisfy

$$\sum_n K_n P_{nn} = 0, \tag{5.20}$$

which can be used to specify P_C in terms of the face pressures once we have determined the second-order gradients.

If the cell tangent directions are orthogonal, and they are used as the unit vectors $\{\hat{e}_n\}$, then the solution to Equation (5.19) is

$$\begin{aligned} P_1 &= (P_E - P_W)/\Delta x, & P_{11} &= 4(P_E + P_W - 2P_C)/\Delta x^2, \\ P_2 &= (P_N - P_S)/\Delta y, & P_{22} &= 4(P_N + P_S - 2P_C)/\Delta y^2, \\ P_3 &= (P_T - P_B)/\Delta z, & P_{33} &= 4(P_T + P_B - 2P_C)/\Delta z^2. \end{aligned} \tag{5.21}$$

The central pressure is explicitly

$$P_C = \frac{(K_{XX}/x^2)(P_E + P_W) + (K_{YY}/y^2)(P_N + P_S) + (K_{ZZ}/z^2)(P_T + P_B)}{2(K_{XX}/x^2 + K_{YY}/y^2 + K_{ZZ}/z^2)}, \tag{5.22}$$

and similarly in two dimensions.

Equations (5.13) and (5.16) each have sufficient degrees of freedom (are complete) with sufficient smoothness to support an upscaling algorithm. Equation (5.13) is the lowest order scheme which supports an upscaling algorithm in a corner-point cell. Because it is build up an octant at a time, it is also preferred for complex cells (PEBI grids), where the generalization to arbitrary number of cell faces is obvious. Equation (5.16) is preferred for regular cells where the solutions (Equations (5.21), (5.22)) make them simple to implement.

5.3. PERMEABILITY AND TRANSMISSIBILITY

Having a new upscaling formulation may be satisfying, but is it useful? Because it works with the most general set of boundary conditions allowed by a finite difference scheme it can be expected to supply an absolute upper bound to estimates of permeability and transmissibility *if* we can derive these quantities from the cell transmissibilities, T_{rs}. Fortunately, we have already developed the necessary techniques in Section 3, when we removed degrees of freedom from the full tensor permeability to derive vector permeability and face transmissibility. As we did there, we ask that the general and the specific models give the same predicted fluxes, in this case being the N_f face fluxes, Q_r. For example, Equation (5.6) is a calculation of Q_r for a triangular element characterized by a permeability tensor.

Based on the construction Equation (5.5) for ∇P, the cell transmissibility for a corner point cell is

$$T_{rs} = -\mathbf{n}_r \bullet \mathsf{K} \bullet \left\{ \sum_{n=1}^{N} \mathbf{t}_{n0}\mathbf{t}_{n0} \right\}^{-1} \bullet \mathbf{t}_{s0}. \tag{5.23}$$

If the cell transmissibility was determined by an upscaling procedure, then to reduce it to a permeability tensor, the determinant

$$0 = \left\| T_{rs} + \mathbf{n}_r \bullet \hat{e}_i K_{ij} \hat{e}_j \bullet \left\{ \sum_{n=1}^{N} \mathbf{t}_{n0}\mathbf{t}_{n0} \right\}^{-1} \bullet \mathbf{t}_{s0} \right\|, \quad i,\, j = 1, \ldots, N, \tag{5.24}$$

must vanish. Here we have written $\mathsf{K} = \sum_{i,\, j=1}^{N} \hat{e}_i K_{ij} \hat{e}_j$ in component form, and the $\{\hat{e}_i\}$ are assumed to be orthonormal. For vector permeability,

$$0 = \left\| T_{rs} + \mathbf{n}_r \bullet \hat{t}_{\ell 0} K_{\ell} \hat{t}_{\ell 0} \bullet \left\{ \sum_{n=1}^{N} \mathbf{t}_{n0}\mathbf{t}_{n0} \right\}^{-1} \bullet \mathbf{t}_{s0} \right\|, \quad \ell = 1, \ldots, N, \tag{5.25}$$

and for face transmissibility

$$0 = \| T_{rs} + \delta_{rs} T_r \|, \tag{5.26}$$

Often, because the row sum of T_{rs} vanishes, the determinant will vanish identically, providing no useful information. If the column sum does not also vanish, then we introduce another condition, Equation (5.9) to build a matrix with both vanishing row and column sum. We may then drop an arbitrary row and column to evaluate the determinants, Equations (5.23)–(5.25). For the determination of permeability, we typically choose Equation (5.20) or (5.22) as this extra condition, as they follow from the local requirements of incompressible flow. As this choice of $\{w_f\}$ depends upon the (unknown) permeability, we have introduced an algebraic nonlinearity into the problem. However, in our experience, it is solved rapidly by iteration.

5.4. DISSIPATION

We are now in a position to return to a claim made in Section 2 – that the dissipation cannot be evaluated in terms of the cell averaged ∇P and $\mathbf{V}$. From explicit representations of $P(\mathbf{x})$ and $\mathbf{V}(\mathbf{x})$ we can evaluate the volume averages of $-\mathbf{V} \bullet \nabla P$, $\mathbf{V}$,

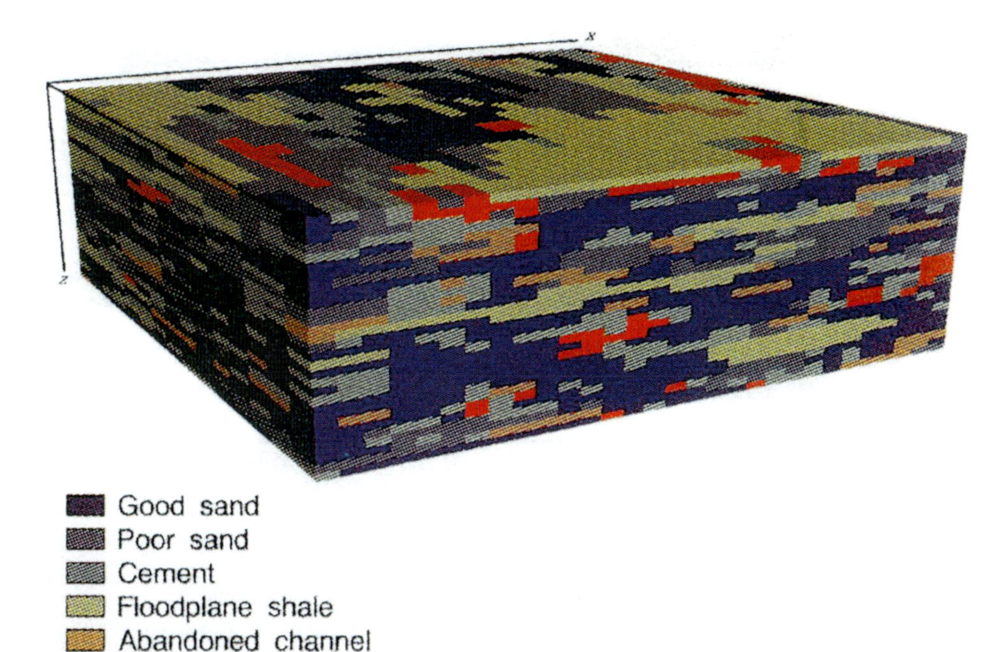

Fig. 7. Sequential indicator simulation of the six lithotypes of a North Sea fluvial reservoir. The large continuous floodplane shales generate large tortuosity and low permeability in the vertical direction.

and $-\nabla P$, and contrast the total dissipation to its estimate obtained from the averages of **V** and $-\nabla P$. Working from the representation Equation (5.16), and assuming that K is symmetric and $\{\hat{e}_n\}$ are its eigenvectors:

$$\begin{aligned} \langle \nabla P \rangle &= \sum_{n=1}^{N} \hat{e}_n (P_n + \langle x_n \rangle P_{nn}), \\ \langle -\mathbf{V} \rangle &= \sum_{n=1}^{N} K_n \hat{e}_n (P_n + \langle x_n \rangle P_{nn}) \end{aligned} \tag{5.27}$$

and

$$\begin{aligned} \langle -\mathbf{V} \bullet \nabla P \rangle &= \sum_{n=1}^{N} K_n \{ (P_n + \langle x_n \rangle P_{nn})^2 + P_{nn}^2 (\langle x_n^2 \rangle - \langle x_n \rangle^2) \} \\ &= \langle -\mathbf{V} \rangle \bullet \langle \nabla P \rangle + \sum_{n=1}^{N} K_n P_{nn}^2 (\langle x_n^2 \rangle - \langle x_n \rangle^2). \end{aligned} \tag{5.28}$$

Hence, the dissipation of flow through the cell is under-represented in terms of the average flux and pressure drops, except when there are no higher-order derivatives

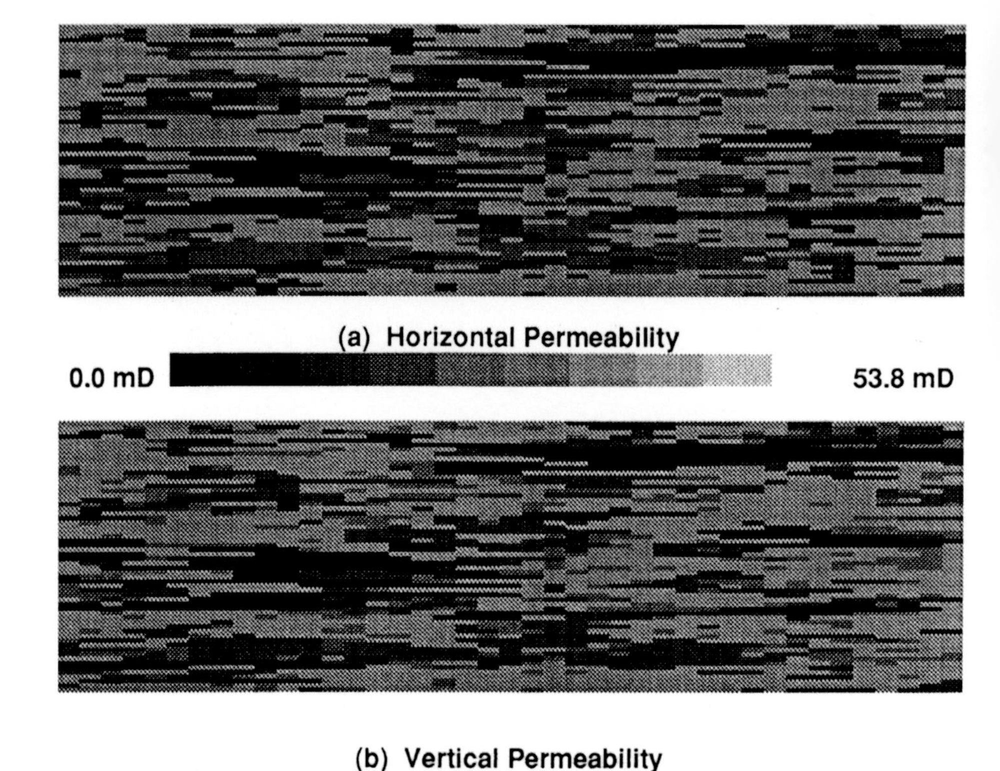

Fig. 8. Horizontal and vertical permeabilities on a two-dimensional (41×60) vertical slice of the reservoir.

in the flow, $P_{nn} = 0$. This demonstrates the claim of Section 2 that dissipation and flux conservative upscaling schemes are not equivalent. An important exception arises for triangular elements, where these higher-order derivatives do vanish. In this case Equation (5.28) demonstrates that the dissipation and flux arguments are equivalent and, hence, the resulting permeability tensor must be symmetric.

5.5. FLUX AND PRESSURE CONTINUITY

Although we may choose eventually to approximate the flux through each face of a cell in terms of a single transmissibility, this is not a full description of the flux. A complete representation, consistent with the order of the finite-element approximation for pressure, is obtained by evaluating total flux and the face-centred pressure on either side of a face, and equating their values. The primary variables are the face-centred pressures, *not* the standard block-centred ones. On corner-point cells, this will give a 7-point scheme in two dimensions, and an 11-point in three. This scheme is applied in two dimensions for the North Sea example to follow. It should be mentioned that an advantage of tetrahedral elements [23, 24] is that

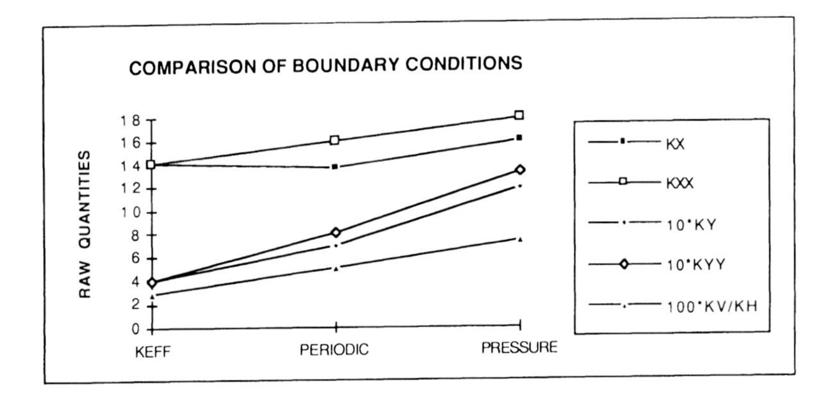

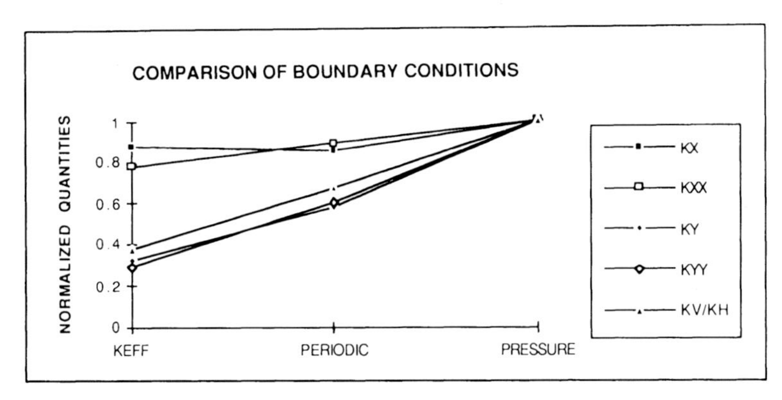

Fig. 9. Influence of boundary conditions on tensor permeability.

the flux continuity approach will give 5 and 7 banded pressures schemes, in two and three dimensions, respectively and, hence, we may use the pressure-solver technology which is readily available.

5.6. NORTH SEA EXAMPLE

We apply the resulting scheme to the North Sea fluvial reservoir example of reference [26]. To review, the laterally continuous flood plane shales are the dominant flow barriers. Figures 7 and 8 show a lithologic description of the reservoir in three dimensions and a permeability mapping in vertical cross-section.

Once the calculations are performed, and a full $2N \times 2N$ cell transmissibility matrix is constructed, then we may restrict the flow according to Equation (5.25) to determine permeability. The results are shown in Figures 9(a) and 9(b), in which we see that the effect on horizontal flow (aligned with the floodplane shales) is $\sim 10\%$ while the vertical flow may be in error by several hundred per cent. The periodic boundary conditions are those of reference [26] and are seen to be intermediate between those of the current study and completely restricted flow.

5.7. DISCUSSION

In this section we have replaced the standard model of upscaling (flow based on a constant pressure gradient) by one which recognises that numerical representations offer degrees of freedom that may be used to capture more of the sub-grid structure than we typically retain in our upscaling algorithms. By using a finite-element representation for pressure, we motivate an upscaling algorithm that gives a *complete* representation of how an averaging region will react under any of a series of slowly varying pressure gradients.

As a complete representation, it offers the most flexible set of upscaling boundary conditions. In Section 3 we recommended upscaling to tensor permeability and then restricting the results to face transmissibility. Now that we have a yet more flexible set of boundary conditions, we recommend upscaling to cell transmissibility first, and only afterwards restricting the results, to face transmissibility. In the North Sea example we show a factor-three variation in vertical permeability depending upon these boundary conditions. As discussed in Section 3, the best approximation is the one that most closely follows the large-scale flow directions, *which are not known*. Nonetheless, by determining absolute upper and lower bounds we can assess the sensitivity of the results to our lack of knowledge. For instance, at the same time as the vertical permeability shows a 300% variation, the horizontal permeability shows only a 10% change.

This Section is summarized in Tables I and II. Table I(b) lists the generalized models of permeability and transmissibility while Table II lists the equation numbers for upscaling and for converting one type of effective property to another.

6. Summary

We have offered a perspective on upscaling of fluid flow based on the simple observation that pressure gradients generate flux, with transport coefficients we may call permeability or transmissibility. Implicit is the recognition that as our representation of the pressure gradient varies, so too must the transport coefficients. They are, in a sense, the duals of each other, matched by their ability to predict the flux. Upscaling can be thought of as moving a slowly varying spatial pressure gradient over a fine structure, and determining the appropriate slowly varying transport coefficients – the coarse grid permeability or transmissibility.

When the pressure gradients are local (differential formulation – Section 2) we have access to a variational approach based on dissipation/ fluctuation. The most important result is a proof for the symmetry of the permeability tensor.

When the pressure gradient is constant in a region (Sections 3 and 4) we have the standard model of permeability and transmissibility. Tensor permeability describes the unconstrained response of the region, while a 'vector' of permeabilities or transmissibilities describe the response with flow restricted into a series of tangent or normal directions. To make the role of directionality more apparent, we have provided a development of the upscaling equations for a general corner-point cell.

We have demonstrated that the principal cause of errors is the choice of boundary conditions which, of necessity, reduce the number of degrees of freedom within the problem. We have also shown how, in the very specialized case of Cartesian renormalization, one way these errors can be estimated and to some extent eliminated (Section 4).

When the pressure gradient is generalized to the degrees of freedom available in a numerical calculation, we must generalize 'permeability' into a quantity we call the cell transmissibility (Section 5). This represents *all* of the ways we may drive fluid through a region, at least within the context of a particular numerical scheme.

We recognize that standard numerical simulators do not provide for these degrees of freedom. Nonetheless, we believe that by upscaling with the most general model first, and then restricting the flow directions and transport parameters later, that we capture more of the possible flow in the upscaling region. This was shown explicitly in Figure 9 where the interaction of the boundary conditions with the domain lead to a 300% variation in effective vertical permeability. To support this approach we have derived a means of converting one representation to another – summarized in Table II.

Triangular elements are very interesting special case. Because of the extreme simplicity of this representation, it has the closest contact with the differential formulation of Darcy's law. Not only is the description equivalent to a permeability tensor, but it even retains the symmetry of the differential form – K must be symmetric.

Although we have only discussed single-phase flow, our motivation has always been the prediction of multiphase flow in porous media: water or gas displacing oil. The emphasis we have placed on the pressure boundary conditions is also required when upscaling multiphase flow, where the effects of cross-flow couple with gravity and the dynamics of the displacement process.

Acknowledgements

The authors would like to thank BP Exploration Operating Company Ltd. for permission to publish this paper, and acknowledge the work of B. Williams and A. Beer (GeoVisual Systems) in testing the corner-point versions of the upscaling algorithms.

References

1. *Reservoir Characterisation*, eds L. W. Lake and H. B. Carroll, Academic Press, 1986.
2. *Reservoir Characterisation II*, eds L. W. Lake, H. B. Carroll and T. C. Wesson, Academic Press, 1991.
3. *Reservoir Characterisation III*, ed B. Linville, Penn Well Publishing, 1993.

4. *North Sea Oil and Gas Reservoirs II*, eds A. T. Buller, E. Berg, O. Hjelmeland, J. Kleppe, O. Torsæter and J. O. Aasen, Graham and Trotman, London, 1990.
5. *North Sea Oil and Gas Reservoirs III*, eds J. O. Aasen, E. Berg, A. T. Buller, O. Hjelmeland, R. M. Holt, J. Kleppe and O. Torsæter, Kluwer Academic, Dordrecht, 1994.
6. *Mathematics in Oil Production*, eds S. F. Edwards and P. R. King, Oxford University Press, 1988.
7. *Mathematics of Oil Recovery*, ed. P. R. King, Oxford University Press, 1992.
8. *2nd Conference on the Mathematics of Oil Recovery*, eds D. Guerillot and O. Guillon, Editions Technip, Paris, 1988.
9. *3nd Conference on the Mathematics of Oil Recovery*, eds M. A. Christie *et al.*, Delft University Press, 1992.
10. Fox. C.: Reservoir Characterisation using expert knowledge, data and statistics, *Oil Field Review*, Jan. 1992.
11. White, C. D., and Horne, R. N.: Computing absolute transmissibility in the presence of fine-scale heterogeneity, paper SPE 16011, *9th SPE Symp. Reservoir Simulation*, San Antonio, Texas (Feb. 1–4, 1987).
12. Indleman, P., and Dagan, G.: Upscaling of permeability of anisotropic heterogeneous formations, Parts 1–3, *Water Resour. Res.* **29** (1993), 917–943.
13. Warren, J.E., and Price, H.S.: Flow in heterogeneous porous media, *SPE J.* Sept. 1961.
14. Begg, S. H., Carter, R. R. and Dranfield, P.: Assigning effective values to simulator gridblock parameters for heterogeneous reservoirs, *SPE RE* Nov. 1989.
15. Holden, L., Hoiberg, J. and Lia, O.: An Estimator for the Effective Permeability, Ref. [8].
16. Wilson, K. G.: The renormalisation group, *Rev. Mod. Phys.* **47** (1975), 773.
17. Kirkpatrick, S.: Models of disordered materials, in *Ill Condensed Matter*, eds R. Balian, R. Maynard and G. Toulouse, North-Holland, Amsterdam, 1979.
18. King, P. R.: The use of renormalisation for calculating effective permeability, *Transport in Porous Media* **4** (1989), 37.
19. King, P. R., Muggeridge, A. H. and Price, W. G.: Renormalisation calculations of immiscible flow, *Transport in Porous Media* **12** (1993), 237.
20. Grindheim, A. O., and Aasen, J. O.: An evaluation of homogenisation techniques for absolute permeability, *Lerkendal Petroleum Engineering Workshop*, Trondheim, 1991.
21. Aharony, A, Hinrichsen, E. L., Hansen, A, Feder, J. Jøssang, T. and Hardy, H. H.: Effective renormalisation group algorithm for transport in oil reservoirs, *Physica A* **177** (1991), 260.
22. Edwards, M. G., and Christie, M. A.: Dynamically Adaptice Godunov Schemes with Renormalisation in Reservoir Simulation, SPE 25268, 1993.
23. Durlofsky, L. J.: Numerical calculation of equivalent grid block permeability tensors for heterogeneous porous media, *Water Resour. Res.* **27** (1991), 699.
24. Durlofsky, L. J.: Modeling Fluid Flow Through Complex Reservoir Beds, *SPE FE* (Dec. 1992) 315–322.
25. Pickup, G. E., Jensen, J. L., Ringrose, P. S. and Sorbie, K. S.: A Method for Calculating Permeability Tensors using Perturbed Boundary Conditions, Ref. [9], 1992.
26. King, M. J.: Application and analysis of a new method for calculating tensor permeability, in *New Developments in Improved Oil Recovery*, ed De Haan, Geological Society Special Publication No. 84, 1975.
27. Ponting, D. K., in Ref. [7], p. 45, 1992.
28. Aziz, K. and Settari, A.: *Petroleum Reservoir Simulation*, Elsevier, Amsterdam, 1979.
29. Ripley, B. D.: *Spatial Statistics*, Wiley, New York, 1989, Ch 4.
30. Malick, K. M., and Hewett, T. A.: Boundary Effects in the Successive Renormalisation of Absolute Permeability, Stanford Center for Reservoir Forcasting (SCRF), Annual Report, May 1994.
31. King, P. R., and Williams, J. K.: Upscaling permeability: Mathematics of renormalization, in *4th Euro. Conf. Mathematics of Oil Recovery*, 1994.